Two week

Apoptosis Methods and Protocols

METHODS IN MOLECULAR BIOLOGY™

John M. Walker, SERIES EDITOR

METHODS IN MOLECULAR BIOLOGY™

Apoptosis Methods and Protocols

Edited by

Hugh J. M. Brady

Institute of Child Health,
University College London,
London, UK

HUMANA PRESS ✳ TOTOWA, NEW JERSEY

© 2004 Humana Press Inc.
999 Riverview Drive, Suite 208
Totowa, New Jersey 07512

humanapress.com

This publication is printed on acid-free paper. ∞
ANSI Z39.48-1984 (American Standards Institute)

Permanence of Paper for Printed Library Materials.

Production Editor: Robin B. Weisberg
Cover illustartion: From Fig. 1 in Chapter 11, "Methods for Culturing Primary Sympathetic Neurons and for Determining Neuronal Viability," by Jonathan Whitfield, Stephen J. Neame, and Jonathan Ham. Cover design by Patricia F. Cleary.

For additional copies, pricing for bulk purchases, and/or information about other Humana titles, contact Humana at the above address or at any of the following numbers: Tel.: 973-256-1699; Fax: 973-256-8341; E-mail: humana@humanapr.com; or visit our Website: www.humanapress.com

Printed in the United States of America. 10 9 8 7 6 5 4 3 2 1

1-59259-812-9 (e-ISBN)

Library of Congress Cataloging-in-Publication Data

Apoptosis methods and protocols / edited by Hugh J. M. Brady.
    p. ; cm. -- (Methods in molecular biology, ISSN 1064-3745 ; 282)
  Includes bibliographical references and index.
  ISBN 0-89603-873-4 (alk. paper)
  1. Apoptosis--Laboratory manuals. 2. Molecular biology--Laboratory manuals.
  [DNLM: 1. DNA Damage--physiology--Laboratory Manuals. 2. Apoptosis--physiology--Laboratory Manuals. QH 465 A644 2004] I. Brady, Hugh J. M. II. Series: Methods in molecular biology (Clifton, N.J.); v. 282.
  QH671.A6585 2004
  571.9'36--dc22

                                                                        2003027914

Preface

The most fundamental question facing each and every cell within an organism is to survive or to die. Cell death is required for normal function; some estimates suggest that as many as one million cells undergo cell death every second in the adult human body. Almost all cells undergoing physiological, or programmed, cell death, independent of cell type, manifest a stereotypic pattern of morphological changes termed apoptosis. Typically, apoptotic cells display shrinkage, membrane blebbing, chromatin condensation, and nuclear fragmentation. The integrity of the cell membrane is not lost during apoptosis and so avoids eliciting the inflammatory response that would have been caused by the spillage of the cell's contents. This is quite in contrast to the loss of cell contents typical of necrosis. The caspases, the family of intracellular cysteine proteases associated with apoptosis, are responsible for the stereotypical morphological changes. Caspases cleave various substrate proteins that act on DNA fragmentation, nuclear envelope integrity, the cytoskeleton, and cell volume regulation. Apoptotic cells are cleared in vivo by the process of phagocytosis, in which specific "phagocytes" move to the site of apoptosis, engulf the dying cells and digest them.

Apoptosis has a central role in many physiological processes, for example, in the immune system. Autoreactive cells are deleted via apoptosis to prevent autoimmunity. At the end of an immune response, activated lymphocytes are removed to maintain homeostasis within the immune system. In addition, immune surveillance involves the use of apoptosis to remove aberrant cells that are damaged or harbor a mutation. When the biochemical pathways regulating apoptosis within the cell are defective in any way, this can make a major contribution to the development of pathological states. For example, insufficient apoptosis can contribute to cancer, autoimmunity, or the persistence of infection. Extensive apoptosis is associated with many diseases, such as heart failure, ischemic injury, and neurodegeneration. Therefore, procedures for detecting apoptosis, quantifying apoptosis, understanding the biochemistry of apoptosis, and identifying the genes and proteins that regulate and carry out apoptosis are crucial techniques in 21st-century biomedical research.

Apoptosis Methods and Protocols introduces the principle of each technique and takes the reader through the procedure in a step-wise, user-friendly fashion. The clear explanations and inclusion of tips learned from many years experience using each technique allow even the novice researcher to successfully carry out procedures as diverse as immunohistochemistry, flow cytometry, kinase activity assays, yeast two-hybrid screening, or the cloning of novel genes by differential expression as related to apoptosis.

Hugh J. M. Brady

Contents

Contributors

RUDI BEYAERT • *Department of Molecular Biomedical Research, Flanders Interuniversity Institute for Biotechnology, Ghent University, Ghent, Belgium*

HUGH J. M. BRADY • *Molecular Hematology and Cancer Biology Unit, Institute of Child Health, University College London, London, UK*

FREDRIK BRYNTESSON • *Molecular Hematology and Cancer Biology Unit, Institute of Child Health, University College London, London, UK*

CARME CAELLES • *Institut de Recerca Biomèdica de Barcelona-Parc Científic de Barcelona, Department of Biochemistry and Molecular Biology, University of Barcelona, Barcelona, Spain*

PHILIP J. COATES • *Division of Pathology and Neurosciences, Ninewells Hospital and Medical School, University of Dundee, Dundee, Scotland, UK*

ANDREW DEVITT • *Centre for Inflammation Research, The University of Edinburgh Medical School, Edinburgh, UK*

RICHARD C. DUKE • *CERES Pharmaceuticals Ltd., Medical Oncology, University of Colorado Health Science Center, and Eleanor Roosevelt Institute for Cancer Research, Denver, CO*

OSKAR S. FRANKFURT • *University of Miami Medical School and Apostain Inc., Miami, FL*

GABRIEL GIL-GÓMEZ • *Institut Municipal d'Investigació Mèdica i Universitat Pompeu Fabra, Barcelona, Spain*

CHRISTOPHER D. GREGORY • *Centre for Inflammation Research, The University of Edinburgh Medical School, Edinburgh, UK*

PETER A. HALL • *Department of Pathology and Cancer Research Centre, Queen's University, Belfast, Northern Ireland, UK*

JONATHAN HAM • *Molecular Hematology and Cancer Biology Unit, Institute of Child Health, University College London, London, UK*

KAREN HEYNINCK • *Department of Molecular Biomedical Research, Flanders Interuniversity Institute for Biotechnology, Ghent University, Ghent, Belgium*

MICHAEL HUBANK • *Molecular Hematology and Cancer Biology Unit, Institute of Child Health, University College London, London, UK*

TERRI KAGAN • *Department of Biology, Queens College and Graduate Center, City University of New York, Flushing, NY*

MARJA KREIKE • *Department of Molecular Biomedical Research, Flanders Interuniversity Institute for Biotechnology, Ghent University, Ghent, Belgium*

GUIDO KROEMER • *CNRS-UMR 8125, Institut Gustave Roussy, Villejuif, France*

SHARAD KUMAR • *Hanson Institute, Institute of Medical and Veterinary Science and Department of Medicine, University of Adelaide, Adelaide, Australia*

JULIE LEBLANC • *Institute for Biological Sciences, National Royal Council, Ottawa, Ontario, Canada*

YANN LEVERRIER • *Laboratory of Molecular and Cellular Biology, Unité de Recherche CNRS/ENS, Lyon, France*

JACQUELINE MARVEL • *Laboratory of Molecular and Cellular Biology, Unité de Recherche CNRS/ENS, Lyon, France*

DEMETRIUS MATASSOV • *Department of Biology, Queens College and Graduate Center, City University of New York, Flushing, NY*

ANNE-LAURE MATHIEU • *Laboratory of Molecular and Cellular Biology, Unité de Recherche CNRS/ENS, Lyon, France*

KIMBERLY MCCALL • *Department of Biology, Boston University, Boston, MA*

DAVID J. MCCONKEY • *Department of Cancer Biology, The University of Texas M.D. Anderson Cancer Center, Houston, TX*

RON MITTLER • *Department of Biochemistry, University of Nevada, Reno NV*

MÓNICA MORALES • *Division of Molecular Biology, Memorial Sloan-Kettering Cancer Center, New York, NY*

STEPHEN J. NEAME • *Eisai London Research Laboratories, University College London, London, UK*

LETA NUTT • *Department of Cancer Biology, The University of Texas M.D. Anderson Cancer Center, Houston, TX*

JEANNE S. PETERSON • *Department of Biology, Boston University, Boston, MA*

JENNIFER REGAN • *Department of Anatomy and Developmental Biology, University College, London, UK*

LUDMILA RIZHSKY • *Department of Botany, Plant Sciences Institute, Iowa State University, Ames, IA*

VICKI SAVE • *Department of Pathology, Addenbrookes Hospital, Cambridge University, Cambridge, UK*

DAVID G. SCHATZ • *Section of Immunobiology, Howard Hughes Medical Institute, Yale University, New Haven, CT*

PETER SCHOTTE • *Department of Molecular Biomedical Research, Flanders Interuniversity Institute for Biotechnology, Ghent University, Ghent, Belgium*

VLADIMIR SHULAEV • *Virginia Bioinformatics Institute, Blacksburg, VA*

MARIANNA SIKORSKA • *Institute for Biological Sciences, National Royal Council, Ottawa, Ontario, Canada*

JOËLLE THOMAS • *Laboratory of Molecular and Cellular Biology, Unité de Recherche CNRS/ENS, Lyon, France*

WIM VAN CRIEKINGE • *Department of Molecular Biomedical Research, Ghent University–VIB, Ghent, Belgium*

SOFIE VAN HUFFEL • *Department of Molecular Biomedical Research, Ghent University–VIB, Ghent, Belgium*

JONATHAN WHITFIELD • *Eisai London Research Laboratories, University College London, London, UK*

OWEN WILLIAMS • *Molecular Hematology and Cancer Biology Unit, Institute of Child Health, University College London, London, UK*

ZAHRA ZAKERI • *Department of Biology, Queens College of City University of New York, Flushing, NY*

NAOUFAL ZAMZAMI • *Centre National de la Recherche Scientifique, UMR 1599, Institut Gustave Roussy, Villejuif, France*

1

Measurement of Apoptosis by DNA Fragmentation

Demetrius Matassov, Terri Kagan, Julie Leblanc, Marianna Sikorska, and Zahra Zakeri

Summary

Classical apoptotic cell death can be defined by certain morphological and biochemical characteristics that distinguish it from other forms of cell death. One such feature, which is a hallmark of apoptosis, is DNA fragmentation. In dying cells, DNA is cleaved by an endonuclease that fragments the chromatin into nucleosomal units, which are multiples of about 180-bp oligomers and appear as a DNA ladder when run on an agarose gel. Here, we present commonly used methods such as conventional agarose gel electrophoresis to analyze fragmented nuclei in cells. The various methods used are dependent on the extent of fragmentation or the amount of fragmented nuclei in a sample. Determining whether a cell exhibits DNA fragmentation can provide information about the type of cell death occurring and the pathways activated in the dying cell.

Key Words: Apoptosis; DNA fragmentation; electrophoresis; pulse-field; agarose gel.

1. Introduction

The methods described in this chapter detect damaged DNA in cells that have fragmented their genome during programmed cell death or apoptosis. Apoptosis is an active regulatory cellular response to certain stimuli, such as loss of trophic factor support (serum deprivation) or exposure to cytotoxic agents (chemotherapeutic

From: *Methods in Molecular Biology, vol. 282: Apoptosis Methods and Protocols*
Edited by: H. J. M. Brady © Humana Press Inc., Totowa, NJ

drugs), that results in a somewhat stereotyped cell death *(1–4)*, including membrane blebbing, nuclear condensation, and the formation of apoptotic bodies. A biochemical hallmark of apoptosis is the cleavage of chromatin into small fragments, including oligonucleosomes, that when seen in electrophoresed gels are described as "DNA ladders" *(5,6)*. DNA fragmentation was first documented by Williamson in 1970 when he observed discrete oligomeric fragments occurring during cell death in primary neonatal liver cultures *(7)*. After that discovery, Williams et al. *(8)* reported the presence of a specific endonuclease activity in cultured Chang liver and Chinese hamster lung cells during cell death *(8,9)*.

DNA fragmentation occurs in several stages. Typically, it commences with the generation of high-molecular-weight (HMW) DNA fragments that are undetectable by conventional agarose gel electrophoresis (CAGE). These large fragments, 50–300 kb in size, reflect the cleavage of interphase chromosomes at the nuclease sensitive sites. These sites are present in the chromatin fiber as a result of its folding into loop (mean size of 50 kb) and rosette (mean size of 300 kb) structures. These DNA fragments can be detected by pulsed-field electrophoresis (PFGE; *10–15*). The large DNA fragments undergo further cleavage as a result of activation of the DNA fragmentation factor *(16)* and Ca^{2+}- and Mg^{2+}-dependent endonucleases. As much as 30% of the chromatin can be cut between nucleosomes at the linker sites *(17)*, producing oligomers that are multiples of about 180 bp (nucleosomal units) which, on agarose gels, appear as characteristic "DNA ladders." Although the appearance of a DNA ladder is a good marker of apoptotic cell death *(17–20)*, it is now well documented that many cells undergoing apoptosis do not degrade their DNA to this extent. Therefore, it should be noted that the formation of the ladder is only the end point of DNA degradation process and it does not represent the full pattern of DNA fragmentation occurring during apoptosis.

There appears to be a major difference between how DNA is cleaved during apoptosis as compared with isolated nuclei incubated with exogenous nucleases *(21,22)*. During apoptotic cell death, the chromatin in cells is not degraded down to the level of single mononucleosomes, as occurs in DNA digested with exogenous nuclease, indicating that not all regions of DNA are nuclease sensitive or that DNA degradation may be specifically turned off at one point *(9)*. Another important difference is the formation of the DNA ends. Irradiated and glucocorticoid-treated apoptotic thymocytes generate small DNA fragments with ends that are exclusively 3' —OH and 5' —P. Endonucleases produce fragmented DNA ends where the bond broken is either at carbon 5 of the sugar, producing a 5' —OH and a 3' —P with the phosphate on carbon 3 of the adjacent sugar, or between the phosphate and carbon 3 to give 3' —OH and 5' —P. Ca^{2+}/Mg^{2+}-dependent endonucleases generate 3' —OH ends whereas acid endonuclease produces a 3' —P end.

Three specific methods, appropriate for the characterization of DNA fragmentation during apoptosis, will be discussed below (*see* **Figs. 1** and **2**). The CAGE protocol, outlined in **Subheading 3.1.**, allows the detection of low-molecular-weight (LMW) DNA ladders *(23)*. A refinement of CAGE, radioactive end labeling, can be used to detect very small quantities of fragmented DNA, as described in **Subheading 3.2.** *(6)*. Large DNA fragments of 50 kB–1 mb, also known as HMW DNA, as well as intact DNA can be detected by PFGE, the protocol for which is given in **Subheading 3.3.** *(10,11)*.

CAGE was first described for cell death by Wyllie *(5)* and subsequently modified in our laboratory. This method has three major steps. First, the fragmented DNA is extracted from the cells. Next, the DNA is purified to remove protein and RNA contaminants that could interfere with the resolution of the DNA ladder. Finally, the DNA is resolved by electrophoresis and is photographed to record the results. Two additional steps are required for Method 2: the incorporation of a radioactive label [^{32}P ddATP] onto the 3'-ends of the fragmented DNA and the detection of that label. Method 3 does not require DNA extraction but relies instead on embedding the cells in agarose plugs. These plugs are run on a field inversion gel electrophoresis (FIGE) Mapper system (Bio-Rad Laboratories), and finally the gels are stained, destained, and photographed. Cells and

METHOD 1 Conventional Agarose Gel Electrophoresis

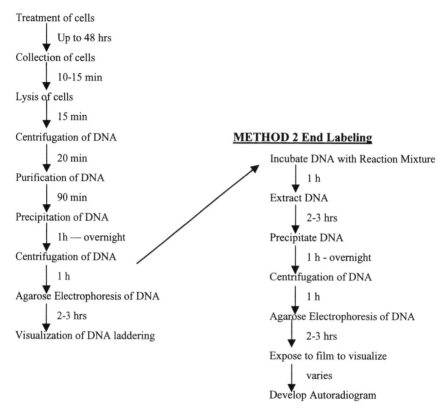

Fig. 1. Outline of general procedures used to detect damaged DNA by CAGE and end labeling techniques.

tissue have to be processed very carefully for PFGE analysis of HMW DNA to eliminate any mechanical or enzymatic damage to the DNA during the processing of the samples.

The methods described above analyze fragmented nuclei in a population of dying cells but not a single cell. There are two techniques, which will be discussed briefly, that can detect single and/or double nicked DNA *in situ*. A rapid, sensitive, and inexpensive technique called single-cell gel electrophoresis or comet assay is used primarily to detect single- or double-strand DNA breaks in cells *(24,25)*. The comet assay assumes that DNA strand

METHOD 3 Pulsed- Field Gel Electrophoresis

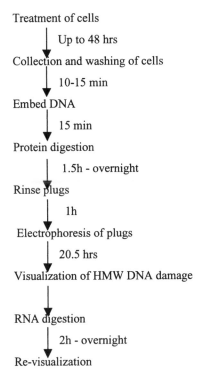

Treatment of cells

Up to 48 hrs

Collection and washing of cells

10-15 min

Embed DNA

15 min

Protein digestion

1.5h - overnight

Rinse plugs

1h

Electrophoresis of plugs

20.5 hrs

Visualization of HMW DNA damage

RNA digestion

2h - overnight

Re-visualization

Fig. 2. Outline of general procedures used to detect damaged DNA by PFGE technique.

breaks are caused as a result of mutagen-induced DNA damage and not from apoptosis mediated nuclear fragmentation *(26)*. The assay is based on the principle that denatured, cleaved DNA fragments migrate from a cell under the influence of an electrical charge applied to the cells, whereas undamaged DNA remains within the cell.

The second technique, called the terminal deoxynucleotidyl transferase (TdT)-mediated dUTP Nick End Labeling assay, can enzymatically detect both double- and single-stranded DNA breaks. The TdT enzyme tags newly generated free 3' OH DNA ends with a modified dUTP nucleotide. This nucleotide can be detected by the use of direct or indirect methods. The direct method of detection uses a dUTP conjugated with a fluorescein molecule whereas the indirect method uses a digoxigenin-11-dUTP nucleotide that is later detected with an antidigoxigenin antibody conjugated with either a peroxidase or fluorescein molecule. The *in situ* labeling can be performed on fixed cells or tissue sections. Because this technique is commercially available by a number of companies (i.e., Roche Diagnostics, cat no. 1684817) the protocol will not be discussed in detail; for additional information, *see* Kagan and Zakeri *(27)*.

2. Materials

2.1. Method 1: Analysis of LMW DNA by CAGE

1. Cell samples: Optimal cell numbers needed for the assay range from 3 to 5×10^6 cells. Any cell line may be examined using this procedure. Alternatively, one may use any type of fresh tissue as long as it is completely homogenized in lysis buffer.
2. Lysis buffer: Make up in 100-mL quantities and store at 4°C for up to 2 mo the following solution: 0.2% Triton X-100; 10 mM Tris-HCl, pH 7.4, 10 mM ethylenediamine tetraacetic acid (EDTA).
3. RNase A: Dissolve 100 mg of RNase A in 10 mL of 15 mM NaCl and 10 mM Tris-HCl, pH 7.4. Boil for 15 min, slowly cool to room temperature, and store in aliquots at –20°C for 2–4 mo.
4. Phenol:chloroform:isoamyl alcohol (25:24:1). Store at 4°C and protect from light. Discard if solution turns reddish.
5. Chloroform:isoamyl alcohol (24:1). Store at 4°C.
6. 5 M NaCl. Store at room temperature.
7. 100% ethanol, ice cold.
8. 70% ethanol. Store at room temperature.
9. TE buffer: 10 mM Tris-HCl, pH 8.0, 10 mM EDTA. Store at room temperature indefinitely.
10. 50× TAE electrophoresis buffer: For 50× stock solution, pH approx 8.5 add 242 g of Tris base, 57.1 mL of glacial acetic acid, 37.2 g of EDTA, and bring to a volume of 1 L by adding dH$_2$O. For 1× working solution (40 mM Tris acetate, 2 mM EDTA) mix 20 mL of stock solution into 980 mL of dH$_2$O. Both solutions can be stored at room temperature indefinitely.

11. 2% agarose: Add 2 g of agarose to 100 mL of 1× TAE running buffer and heat in microwave until dissolved. Store at room temperature.

12. Phosphate-buffered saline (PBS), 10× and 1×

10× stock solution, 1 L	1× working solution, pH 7.4
80 g NaCl	137 mM NaCl
2 g KCl	2.7 mM KCl
11.5 g $Na_2HPO_4 \cdot 7H_2O$	4.3 mM $Na_2HPO_4 \cdot 7H_2O$
2 g KH_2PO_4	1.4 mM KH_2PO_4
pH adjusted to 7.4	

Store at room temperature. 1× PBS solution can become contaminated by microorganisms.

13. 10× loading buffer: Prepare concentrated stock solution by adding the following reagents at the indicated final concentrations: 20% (w/v) Ficoll 400, 0.1 M EDTA, 1.0% sodium dodecyl sulfate (SDS), 0.25% (w/v) bromphenol blue, 0.25% (w/v) xylene cyanol (optional; runs approx 50% as fast as bromphenol blue and can interfere with visualization of bands of moderate molecular weight, but helpful for monitoring very long runs) in dH_2O. Store at room temperature indefinitely.

14. Ethidium bromide stock solution (0.5 mg/mL): Dissolve 50 mg of ethidium bromide in 100 mL of dH_2O. Use diluted 1:1000. Store at 4°C and protect from light. Caution: ethidium bromide is a potential carcinogen. Wear gloves when handling.

2.2. Method 2: Radioactive End Labeling of LMW DNA

1. Isolated LMW DNA.
2. 5× TdT labeling buffer (Boehringer Mannheim).
3. 25 mM cobalt chloride (Boehringer Mannheim).
4. ^{32}P ddATP (Amersham, 10 mCi/mL).
5. TdT: 5 units/μL, in potassium cacodylate solution, pH 7.2, with glycerol, 50% (v/v) (Boehringer Mannheim).
6. 500 mM EDTA, pH 8.0
7. Phenol:chloroform:isoamyl alcohol (25:24:1). Store at 4°C and protect from light.
8. 3 M sodium acetate
9. Glycogen (20 mg/mL) in dH_2O
10. TE buffer, pH 8.0 as in **Subheading 2.1.**

2.2. Method 3: Analysis of HMW DNA Fragments by PFGE

1. Nuclear buffer (NB): 15 mM Tris-HCl, pH 7.4, 1 mM ethylene-bis(oxyethylenenitrilo)tetraacetic acid, 2 mM EDTA, 0.5 mM spermidine, 0.15 mM spermine, 60 mM KCl, 15 mM NaCl. Store at 4°C for up to 2 mo.
2. TE Buffer as in **Subheading 2.1.**, adjusted to pH 8.5.
3. 1.5% embedding agarose: 0.15 g of low melting point agarose in 10 mL of NB. Heat in microwave to dissolve and store at 4°C. Melt in microwave and cool to 37°C before use.
4. Proteinase K (20 mg/mL): Dissolve 20 mg in 1 mL of dH$_2$O. Store aliquots at –20°C.
5. TEEN buffer: 10 mM NaCl, 10 mM Tris-HCl, pH 9.5, 25 mM EDTA, 5 mM ethylenebis(oxyethylenenitrilo)tetraacetic acid.
6. 10% SDS: 10 g SDS in 100 mL of dH$_2$O
7. 20× TBE: 0.89 M Tris base, 0.89 M boric acid, 25 mM EDTA. Use diluted to 1× TBE.
8. 0.8% agarose gel: 2.4 g of agarose in 300 mL of 1× TBE. Heat in microwave to dissolve, cool to 60°C, pour into a gel-casting tray, and insert comb. Let set 30 min at room temperature.
9. Ethidium bromide stock solution (10 mg/mL): dissolve ethidium bromide in 0.8% low melting point agarose in 0.5× TBE. Melt in microwave and cool to 37°C before use.
10. Miniplug casting mold, horizontal gel chamber, and FIGE Mapping System (Bio-Rad).

3. Methods

3.1. Detection of Fragmented DNA by Agarose Gel Electrophoresis

3.1.1. Extraction and Purification of LMW DNA

1. Plate and treat the desired cell line at a cell density of 3–5 × 10^6 cells (*see* **Note 1**).
2. Collection for adherent cells: Gently scrape cells and place cells and supernatant in 15-mL Falcon tube on ice. Centrifuge at 1000g for 5 min at 4°C. Aspirate off old media, resuspend cell pellet in 1 mL of ice-cold 1× PBS, and transfer to a 1.5-mL microcentrifuge tube. Col-

lection for suspension cells: Place media and cells in a 15-mL Falcon tube on ice. Centrifuge at 1000*g* for 5 min at 4°C. Aspirate media, resuspend pellet in 1 mL of ice-cold 1× PBS, and transfer to a 1.5-mL microcentrifuge tube (*see* **Note 2**).

3. Centrifuge sample at 1000*g* for 1 min at 4°C. Aspirate 1× PBS wash and lyse cells by adding 500–600 µL of the lysis buffer. Incubate on ice for 15 min (*see* **Notes 3** and **4**).

4. Centrifuge at 12,000*g* at 4°C for 20 min.

5. Transfer the supernatant, which contains the LMW DNA, into a new tube, add RNase A to a final concentration of 100 µg/mL, and mix completely by flicking the tube (*see* **Note 5**).

6. Incubate at 37°C for 1 h (*see* **Note 6**).

7. Extract by adding an equal volume (500–600 µL) of phenol: chloroform:isoamyl alcohol (25:24:1) to the sample and vortexing for a few seconds to properly mix the solutions. Spin tube at high speed for 3 min at room temperature. Carefully remove the top (aqueous) phase and transfer to a clean tube. Back extract the bottom phenol phase by adding 100 µL of the lysis buffer and vortex to mix. Respin tube at high speed for 3 min at room temperature and combine new upper phase with the previous aqueous phase (*see* **Note 7**).

8. Repeat **step 7**.

9. Add 500 µL of chloroform:isoamyl alcohol (24:1) to the sample and briefly mix by vortexing. Spin tube at high speed for 3 min at room temperature. Remove the top aqueous layer and place in a new tube.

10. Precipitate the DNA by adding 25–30 µL of 5 *M* NaCl to a final concentration of 300 m*M* and add 2–2.5 volume of ice-cold 100% ethanol. Leave overnight at –20°C (*see* **Note 8**).

11. Spin sample at 12,000*g* for 30 min at room temperature. Carefully aspirate the ethanol off and wash the DNA pellet with 1 mL of 70% ethanol. Dislodge the pellet by inverting several times so that the ethanol can remove any excess salts.

12. Spin sample at 12,000*g* for 20 min at room temperature. The pellet may be loose so carefully and completely remove the ethanol wash.

13. Dry the pellet by placing it in a Speedvac evaporator and spin for a couple of minutes. The time will vary depending on how much ethanol was left during the 70% ethanol wash.

3.1.2 Agarose Gel Electrophoresis and Analysis of Fragmented DNA

1. Resuspend the dried DNA pellet by adding 18 µL of TE buffer and flicking the tube a few times.
2. Incubate in water bath for 45°C for 15 min.
3. Put DNA samples on ice and add 2 µL of 10× loading buffer. Microcentrifuge the samples to get all the liquid that may have collected on the sidewalls of the tube.
4. Melt 20 mL of 2% agarose in the microwave. Ethidium bromide can be added to the gel or electrophoresis buffer at 0.5 µg/mL. Swirl to mix the ethidium bromide and pour the agarose into the gel-casting tray. Place gel comb to form the wells.
5. After the gel has hardened (20–30 min), remove the gel comb and place gel into an electrophoresis tank containing sufficient 1× TAE electrophoresis buffer to cover the gel approx 1 mm.
6. Gently aliquot the individual DNA samples into each well. Attach the leads so that the DNA migrates to the anode or positive lead, and electrophorese at 1 to 10 V/cm of gel (*see* **Notes 9** and **10**).
7. Turn off the power supply when the bromphenol blue dye has migrated two thirds of the way down the gel (bromphenol blue comigrates with approx 0.5-kb fragments).
8. Photograph a stained gel directly on an UV transilluminator or first stain gel with 0.5 µg/mL ethidium bromide 10 to 30 min, destaining 30 min in water, if necessary (*see* **Notes 11–14**).

3.2. Radioactive End Labeling DNA Fragmentation Analysis

1. Collect the LMW DNA as described in **Subheading 3.1.1.**
2. Resuspend the dried DNA pellet by adding 20 µL of TE buffer and flicking the tube a few times.
3. Make reaction mixture on ice by adding the following to a Microfuge tube, yielding a final volume of 50 µL per DNA sample: 10 µL of 5× TdT labeling buffer, 5 µL of $CoCl_2$ solution, 20 µL of DNA sample, 0.5 µL of ^{32}P ddATP, 5 µL of terminal transferase (25 units), and 10.5 µL of dH_2O.
4. Incubate reaction mixture at 37°C for 1 h.
5. Stop the reaction by adding 5 µL of EDTA to each reaction tube and heating at 70°C for 10 min.

6. Extract once with 600 µL of phenol:chloroform:isoamyl alcohol (25:24:1) and vortex to mix properly. Centrifuge sample at high speed and carefully remove upper layer to a clean tube. Back extract the phenol to recover as much of the DNA as possible.

7. Precipitate the DNA overnight at –20°C with 1:10 vol sodium acetate, 1 µL of glycogen, and 2.5 vol 100% ethanol.

8. Spin sample at 12,000g for 30 min at room temperature. Carefully aspirate the ethanol off and wash the DNA pellet with 1 mL of 70% ethanol.

9. Spin sample at 12,000g for 20 min at room temperature. The pellet may be loose so carefully and completely remove the ethanol wash.

10. Dry the pellet by placing it in a Speedvac evaporator and spin for a couple of minutes. The time will vary depending on how much ethanol was left during the 70% ethanol wash.

11. Resuspend DNA pellet in 30 µL of TE buffer

12. Use 15 µL of DNA to run a 2% agarose gel

13. Dry agarose gel at 60°C for 4 h and expose gel to film overnight.

3.3. Analysis of HMW DNA Fragments by PFGE

1. Obtain samples as described above in **Subheading 3.1.1., steps 1** and **2**.

2. Resuspend samples completely in 1 mL of ice-cold NB.

3. Spin samples at 1000g for 5 min at 4°C.

4. Resuspend cells in 40 µL of NB and mix with 40 µL of 1.5% embedding agarose containing 1 µL of proteinase K (20 mg/mL).

5. Pipet the mixture into the well of a miniplug casting mold and allow to set for 10 min at 4°C.

6. Eject the plugs into 1.5-mL Microfuge tubes containing 300 µL of TEEN buffer, 15 µL of 10% SDS, and 2 µL of proteinase K (20 mg/mL).

7. Incubate for 1.5 h overnight at 37°C on a rotator.

8. Rinse plugs in cold TE buffer for 30 min. Repeat.

9. Place plugs into the wells of a 0.8% agarose gel and seal in place with molten 0.8% low melting point agarose prepared in 0.5× TBE (*see* **Note 15**).

10. Let set 5 min at 4°C.

11. Run the gel using a horizontal gel chamber and a FIGE Mapper System. Electrophoresis is performed using the following settings: 280 V in the forward direction and 90 V in the reverse direction with

an initial switch time of 0.3 s forward and 0.5 s reverse with nonlinear ramping to 35 s forward and 60 s reverse for a total run time of 20.5 h.

12. Running buffer is recirculated through a Lauda bath set to 14°C.

13. Stain the gel with ethidium bromide for 30 min, destain 30 min to several hours in dH$_2$O, and photograph on a UV transilluminator (*see* **Notes 16** and **17**).

14. Remove any RNA from the gel by soaking it in running buffer containing 1 μg/mL of RNaseA for 2 h at room temperature or overnight at 4°C.

15. Rephotograph.

4. Notes

1. Do not plate fewer than 5×10^5 cells because DNA will not be detectable by photography in ethidium bromide-stained gels and no more than 5×10^6 cells to avoid the problem of handling large amounts of DNA, which is insoluble. A limitation of this procedure is that it requires a large number of cells to have proper levels of DNA laddering. Alternatively, you may use any type of tissue as long as it is completely homogenized in lysis buffer.

2. Scraping of some adherent cells, such as Madin–Darby canine kidney cells, may result in damaged outer plasma membranes and vital dyes, such as trypan blue, will readily enter the cell, giving false viability data. Another problem is that some cell lines, such as mouse embryo fibroblast cells, tend to clump while scraping, leading to poor lysis and low yield of DNA. These potential problems can be avoided by mildly trypsinizing the cells. Remove old media, place in 15-mL Falcon tubes on ice and wash with ice-cold 1× PBS to remove any serum. Remove wash, add to Falcon tubes and add 400–500 μL of a 2.5% trypsin solution and incubate at 37°C for several minutes (time depends on how adherent the cells are to the plate). Do not trypsinize too long because this may also result in cellular damage; therefore, add the soybean trypsin inhibitor (1 μg/mL) to stop trypsin activity. Wash detached cells with 1× PBS and place in a Falcon tube. Follow the remaining portion of **step 2** in **Subheading 3.**

3. Do not centrifuge the cells too long because it will be difficult to resuspend and lyse the cells. Avoid vigorous pipetting to break cell clumps during lysis because such actions contribute to poor DNA laddering. Samples may be incubated longer to ensure complete lysis of the cells.

4. DNA laddering can be detected by this technique in fresh tissue by lysing the homogenized tissue overnight at 37°C in PBS containing SDS and proteinase K (10 mM Tris-HCl, pH 7.8, 5 mM EDTA, 0.5% SDS, and 0.6 μg/μL proteinase K). However, frozen tissue samples should not be used because the freezing/thawing process leads to additional DNA fragmentation.

5. The pellet contains the HMW DNA and can usually be discarded because LMW DNA is generally run on a CAGE gel. However, when no DNA is evident after a gel has run, and there is a question of whether the DNA has been lost during processing or whether there even was DNA present in the sample, the pellet containing the HMW DNA can be run on a gel as follows: Resuspend the pellet in 500 μL of sarcosyl buffer (10 mM EDTA, 100 μg/mL proteinase K, 0.5% N-lauroyl sarcosine sodium salt, 50 mM Tris) and RNase to a final concentration of 100 μg/μL at 50°C for 12–16 h. Extract the DNA with phenol:chloroform as described above and then precipitate with 300 mM NaCl and 2.5 vol ethanol at 4°C overnight. The DNA can then be spooled into a new tube, suspended in TE buffer, and run on a 0.7% agarose gel.

6. The RNase incubation may be longer but should not be skipped because an RNA band may interfere with visualizing the DNA laddering.

7. It is important to obtain pure DNA because contaminants like protein can block DNA from leaving the wells during electrophoresis. Proper mixing of the phenol with the DNA is required to ensure efficient purification but excessive mixing may lead to shearing of the DNA.

8. Ice-cold 100% isopropanol may also be used to precipitate DNA, allowing larger starting volumes during precipitation. Equal volumes of isopropanol should be used in precipitating the DNA. Instead of precipitating the DNA overnight, the samples can be placed in a dry ice/ethanol slurry for 1 h, but it is recommended to do the overnight incubation if time permits.

9. During electrophoresis, a sharp red/orange band may appear at the edge of the well, which indicates large amounts of DNA (the red/orange color is from the ethidium bromide). Usually, the red/orange band means that sufficient DNA has been extracted and will show the typical DNA laddering pattern. However, the quality of the laddering will be evident only after the complete run. Lanes that do not show the band may still have DNA laddering but will be

Lanes   1   2   3   4   5

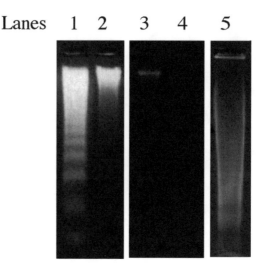

Fig. 3. CAGE gels illustrating different possible outcomes. Lane 1 exhibits good DNA laddering. Lane 2 contains unresolved HMW DNA (*see* **Note 14**). The DNA in lane 3 appears unfragmented. Lane 4 can be interpreted in several different ways: 1) There is no fragmentation in the sample, and the cells are completely whole and healthy; 2) the DNA was so large that it could not penetrate the gel and was lost (*see* **Note 18**); 3) the DNA was lost completely during processing (*see* **Note 5**). The DNA in lane 5 is smeared, which is the result of uncontrolled and complete DNA lysis either during nonapoptotic death or as a result of mishandling of DNA during processing (*see* **Note 13**).

faint. The band is a good marker to see whether you will get any laddering, because samples that should not show laddering will not have the band (**Fig. 3**).

10. A positive control for DNA fragmentation should also be included during the analysis. Some inducers of cell death include serum deprivation, ethanol, tumor necrosis factor-α, anti-CD95 antibody, and serum or growth factor withdrawal. U937 or HeLa cell lines can be used and treated with 100% ethanol to a final concentration between 2 and 2.5%. The ethanol is added directly to the media and samples are incubated for 24–36 h.

11. A large band at the top of the gel may be evident after electrophoresis. This band represents large semifragmented pieces of DNA and indicates incomplete apoptotic fragmentation in the sample material.

12. In some instances, electrophoresis of total cellular DNA is effective and will result in a good fragmentation ladder. Nevertheless, we prefer to extract only the LMW DNA for electrophoresis. Because cells, whether in tissue or in culture, are not homogeneous in the kinetics of their death, separating the low mass DNA from the total DNA can greatly increase the sensitivity of this assay.

13. When the DNA of a cell is completely fragmented, the DNA appears as a smear. This may represent either a necrotic event involving complete lysis of the cell and its contents, or DNase activity and contamination in the extraction process.

14. When DNA remains near the top of a CAGE gel, it is a sign that intact and large fragment DNA is present and PFGE is recommended. Fragments greater than 30 kb cannot be resolved by CAGE.

15. Size markers should be run in the gel alongside samples. Two markers often used for conventional agarose gels are the 123-bp ladder and a *Hind*III digest of λ DNA (Gibco-BRL). Three sets of standards, with overlapping size ranges, are recommended for pulsed-field gels: yeast chromosomes (225 kb–1.9 mb), polymerized λ phage DNA (50 kb–1 mb), and low range marker (0.1 kb–200 kb; New England Biolabs).

16. Gel documentation can also be conducted using a CCD camera coupled to a thermal printer and/or a data acquisition board and computer.

17. Nonspecific fluorescence can be removed by increasing the destaining time.

18. To prevent loss of DNA from lack of penetration, the gel may be loaded while a low voltage is applied. The buffer should not completely cover the gel when loading in this manner.

Acknowledgments

We thank Richard Lockshin for critical reading of the manuscript. We also thank N. Karasavvas and T. Latham for their roles in developing the different methods described in this work. This work was supported by the National Institute of Health Grant 413-48-0001.

References

1. Clarke, P. G. H. (1990) Developmental cell death: morphological diversity and multiple mechanisms. *Anat. Embryol.* **181,** 195–213.

2. Steller, H. (1995) Mechanisms and genes of cellular suicide. *Science* **267**, 1445–1449.
3. Thompson, C. B. (1995) Apoptosis in the pathogenesis and treatment of disease. *Science* **267**, 1456–1462.
4. Vaux, D. L. (1993) Toward an understanding of the molecular mechanisms of cell death. *Proc. Natl. Acad. Sci. USA* **90**, 786–789.
5. Wyllie, A. H. (1980). Glucocorticoid induced thymocyte apoptosis is associated with endogenous endonuclease activation. *Nature* **284**, 555–556.
6. Zakeri, Z. F., Quaglino, D., Latham, T., and Lockshin, R. A. (1993) Delayed internucleosomal DNA fragmentation in programmed cell death. *FASEB J.* **7**, 470–478.
7. Williamson, R. (1970) Properties of rapidly labeled deoxyribonucleic acid fragments isolated from the cytoplasm of primary cultures of embryonic mouse liver cells. *J. Mol. Biol.* **51**, 157–168.
8. Williams, J. R., Little, J. B., and Shipley, W. U. (1974) Association of mammalian cell death with a specific endonucleolytic degradation of DNA. *Nature* **252**, 754–756.
9. Sikorska, M., and Walker, P. R. (1998) Endonuclease activities and apoptosis, in *When Cells Die* (Lockshin, R. A., Zakeri, Z., Tilly, J. L., eds.), Wiley-Liss, Inc., New York, NY, pp. 211–242.
10. Walker, P. R., Leblanc, J., Smith, B., Pandey, S., and Sikorska, M. (1999) Detection of DNA fragmentation and endonucleases in apoptosis. *Methods* **17**, 329–338.
11. Walker, P. R., Kokileva, L., LeBlanc, J., and Sikorska, M. (1993) Detection of the initial stages of DNA fragmentation apoptosis. *BioTechniques* **15**, 1032–1040.
12. Walker, P. R., Weaver, V. M., Lach, B., LeBlanc, J., and Sikorska, M. (1994) Endonuclease activities associated with HMW and internucleosomal DNA fragmentation in apoptosis. *Exp. Cell Res.* **213**, 100–106.
13. Walker, P. R., and Sikorska, M. (1994) Endonuclease activities, chromatin structure and DNA degradation in apoptosis. *Biochem. Cell Biol.* **72**, 615–623.
14. Jochová, J., Quaglino, D., Zakeri, Z., Woo, K., Sikorska, M., Weaver, V., and Lockshin, R. A. (1997) Protein synthesis, DNA degradation, and morphological changes during programmed cell death in labial glands of Manduca sexta. *Devel. Gen.* **21**, 249–257.
15. Brown, D. G., Sun, X. M., and Cohen, G. M. (1993) Dexamethasone-induced apoptosis involves cleavage of DNA to large frag-

ments prior to internucleosomal fragmentation. *J. Biol. Chem.* **268,** 3037–3039.

16. Liu, X., Zou, H., Slaughter, C., and Wang, X. (1997) DFF, a heterodimeric protein that functions downstream of caspase-3 to trigger DNA fragmentation during apoptosis. *Cell* **89,** 175–184.

17. Arends, M. J., Morris, R. G., and Wyllie, A. H. (1990) Apoptosis: the role of the endonuclease. *Am. J. Pathol.* **136,** 593–608.

18. Montague, J. W., Bortner, C. D., Hughes, F. M. Jr., and Cidlowski, J. A. (1999) A necessary role for reduced intracellular potassium during the DNA degradation phase of apoptosis. *Steroids* **64,** 563–569.

19. Bursch, W., Kleine, L., and Tenniswood, M. (1990) The biochemistry of cell death by apoptosis. *Biochem. Cell Biol.* **68,** 1071–1074.

20. Gavrieli, Y., Sherman, Y., and Ben-Sasson, S. A. (1992) Identification of programmed cell death in situ via specific labeling of nuclear DNA fragmentation. *J. Cell Biol.* **119,** 493–501.

21. Umansky, S. R., Beletsky, I. P., Korol, B. A., Lichtenstein, A. V., and Neilpovich, P. A. (1988) Molecular mechanisms of DNA degradation in dying rodent thymocytes. *Mol. Cell Biol.* **7,** 221–228.

22. Beletsky, I. P., Matyasova, J., Nikonova, L. V., Skalka, M., and Umansky, S. R. (1989) On the role of Ca, Mg-dependent nuclease in the postirradiation degradation of chromatin in lymphoid tissues. *Gen. Physiol. Biophys.* **8,** 381–398.

23. Karasavvas, N., Erukulla, R. K., Bittman, R., Lockshin, R., and Zakeri, Z. (1996) Stereospecific induction of apoptosis in U937 cells by N-octanoyl-sphingosine stereoisomers and N-octyl-sphingosine. The ceramide amide group is not required for apoptosis. *Eur. J. Biochem.* **236,** 729–737.

24. Olive, P. L., Banath, J. P., and Durand, R. E. (1990) Heterogeneity in radiation-induced DNA damage and repair in tumor and normal cells measured using the "comet" assay. *Radiat Res.* **122,** 86–94.

25. Lee, R. F., and Steinert, S. (2003) Use of the single cell gel electrophoresis/comet assay for detecting DNA damage in aquatic (marine and freshwater) animals. *Mutat Res.* **544,** 43–64.

26. Lee, R. F., and Steinert, S. (2003) Use of the single cell gel electrophoresis/comet assay for detecting DNA damage in aquatic (marine and freshwater) animals. *Mutat. Res.* **544,** 43–64.

27. Kagan, T., and Zakeri, Z. (1998) Detection of apoptotic cells in the nervous system, in *Methods in Molecular Medicine,* Vol. 22, *Neurodegeneration Methods and Protocols* (Harry, J., ed.), Humana Press, Totowa, NJ, pp. 105–124.

2

Measurement of Caspase Activity in Cells Undergoing Apoptosis

Sharad Kumar

Summary

Cysteine proteases of the caspase family play key roles in the execution of apoptosis and in the maturation of proinflammatory cytokines. During apoptosis signaling, the latent forms of caspase precursors undergo rapid proteolytic processing and activation. Thus, the measurement of caspase activation provides a quick and convenient mean to assess apoptosis. This chapter outlines the various commonly used assays for determining caspase activity in cultured cells or tissue extracts.

Key Words: Caspase activation; synthetic peptides; immunoblotting; electrophoresis; spectrophotometer.

1. Introduction

Caspases (from **c**ysteine **aspases**) are a group of cysteine proteases that cleave their substrates after an aspartate residue *(1–5)*. These proteases are the mammalian homologues of the *Caenorhabditis elegans* death protease CED-3. There are two major functions assigned to caspases. Some caspases, such as caspase-1, -4, -5, and -11, are primarily involved in the processing and activation of proinflammatory cytokines, whereas others, including caspase-2, -3, -6, -7, -8, -9, and -10, have been implicated in the effector phase of

From: *Methods in Molecular Biology, vol. 282: Apoptosis Methods and Protocols*
Edited by: H. J. M. Brady © Humana Press Inc., Totowa, NJ

apoptosis *(1–6)*. Caspases are normally present as inactive precursors in cells. Upon receiving an apoptotic signal, the proforms (zymogens) of caspases undergo proteolytic processing to generate active enzyme *(1–5)*. The structural studies on active caspase-1 and caspase-3 predict that the mature enzymes have a heterotetrameric configuration, composed of two heterodimers derived from two precursor molecules *(7–10)*. In addition to the regions that give rise to two subunits, procaspases contain amino terminal prodomains of varying lengths. Based on the length of prodomain, caspases can be divided into two groups: class I, which contain a relatively long prodomain, and class II, which contain a short prodomain *(11,12)*. From recent studies it has become apparent that the long prodomains in many class I caspases consist of specific protein–protein interaction motifs, such as the caspase recruitment domain and the death effector domain, which play a crucial role in caspase activation. These motifs in the prodomains of class I caspases seem to serve two functions; mediate oligomerization of the procaspase molecules and/or help the recruitment of caspase precursors to specific death adaptor complexes. The clustering of class I procaspase molecules results in caspase activation by autocatalysis by mechanisms that are not entirely understood at present. Class II caspases, which lack a long prodomains, also lack the ability to self-activate and appear to require cleavage by activated class I caspases. For this reason, class I caspases are also referred to as initiator or upstream caspases and class II caspases as executioner or downstream caspases. The activation of class I caspases is of fundamental importance in cell death commitment and, hence, substantial efforts have been devoted to the understanding of mechanisms that underlie the activation of class I caspases *(11,12)*.

Induction of apoptosis is almost always associated with the activation of caspases; therefore, measurement of caspase activity is a convenient way to assess whether the cells are undergoing apoptosis. There are several ways to measure caspase activation. Most common ones, described here, involve use of chromogenic or fluorogenic peptide substrates that release the chromogen or fluorescent tag upon cleavage by a caspase. Other qualitative methods include

Table 1
A List of Commonly Used Synthetic Peptide Substrates for Caspases

Caspase	Optimal substrate	Other substrates
Caspase-1	WEHD	YVAD
Caspase-2	VDVAD	DEVD
Caspase-3	DEVD	VDVAD
Caspase-4	WEHD	YVAD
Caspase-5	WEHD	YVAD
Caspase-6	VEID, VEHD	
Caspase-7	DEVD	VDVAD
Caspase-8	LETD	VEID, DEVD
Caspase-9	LEHD	
Caspase-10	LETD	VEID, DEVD

The peptide substrates usually have an Ac- or z-amino terminal blocking group and either AFC, AMC, or pNA reporter at the carboxyl terminus. The optimal substrates are based on in vitro cleavage specificities determined by screening peptide combinatorial libraries using recombinant caspases expressed in *E. coli* *(4,13,14)*. Alternative substrates that can also be used for caspase assays are listed in the third column.

monitoring the cleavage of in vitro synthesized ^{35}S-labeled caspase substrates or measuring the cleavage of endogenous caspase substrates by immunoblotting using specific antibodies. In this chapter, all three techniques are described.

The most direct and quantitative method for measuring caspase activity is by using synthetic peptide substrates. There are 13 mammalian caspases described. Optimal substrate specificities for many of these have been determined using peptide combinatorial libraries *(13,14)*. The minimum substrate required for a caspase is usually a tetrapeptide sequence with an aspartate residue in P_1 position and variable P_2 to P_4 residues based on cleavage specificity of individual caspases. In some cases, such as caspase-2 and the *Drosophila* caspase DRONC, a P_5 residue greatly enhances substrate cleavage *(14–16)*. The most commonly used substrates that are commercially available are listed in **Table 1**. However, it must be noted that in crude cell extracts containing many caspases, it is not possible to distinguish which caspases are contributing to activity by using substrates listed in **Table 1** because most of the commonly used caspase substrates can be cleaved by more than one caspase, albeit at different efficiencies *(13,14)*. Furthermore, the abundance of individual caspases in a cell type can vary greatly; therefore, the relative contribution of a single caspase to substrate cleavage is always difficult to assess.

2. Materials

1. Cells and tissue samples: Cells and tissue samples in which caspase activities are to be determined are to be supplied by the investigator. For a positive control, mammalian cell lines treated with apoptosis-inducing agents can be used. As a guide, extracts prepared from Jurkat cells treated for 2 h with 200 ng/mL of an anti-Fas antibody (e.g., from Upstate Biotechnology) or for 4 h with 40 μM etoposide, will show significant levels of caspase activity on IETD, DEVD, and VDVAD substrates. Extracts from treated cells can be prepared as described below in **Subheading 3.1.1.** (*see* **Notes 1–3**).

2. Caspase substrates: The fluorogenic substrates, with either acetyl-(Ac) or benzyloxycarbonyl- (z) blocking group and 7-amino-4-trifluoromethylcoumarin (AFC) or 7-amino-4-methylcoumarin (AMC) reporters, and colorimetric substrates, with *p*-nitroanilide (pNA) reporter, are available from various commercial sources. Two of the largest suppliers are Enzyme Systems Products Inc. (United States) and Bachem (Switzerland), but many commonly used caspase substrates and inhibitors can now be bought from numerous suppliers. In our experience, fluorogenic assays are far more (50- to 100-fold) sensitive than the colorimetric assays. This may be an important consideration when there is a limited availability of starting material (cells or tissue sample). AMC and AFC substrates can be stored at –20°C as 5- to 10-mM stock solution in dimethyl formamide for 1–2 yr. Dissolve pNA substrates at 20 mM in dimethyl formamide and store the same as one would AMC and AFC substrates.

3. Spectrometers: For the measurement of fluorescence, a luminescence spectrometer, such as Perkin-Elmer LS50B fluorometer, preferably equipped with a thermostated plate reader, is required. If using pNA colorimetric substrates, a spectrophotometer, preferably equipped with a thermostated cuvette or plate holder, is required.

4. Caspase assay buffers: Prepare caspase assays buffer containing 100 mM 4-(2-hydroxyethyl)-1-piperazineethanesulphonic acid, pH 7.0; 10% sucrose; 0.1% 3-[(3-cholamidopropyl)dimethylam-monio]-1-propanesulfonate; and 10 mM dithiothreitol (DTT) and store in aliquots at –20°C. Alternatively, assays buffer without DTT can be stored at room temperature for several months and DTT added to 10 mM from a fresh 1 M stock as required (*see* **Note 1**).

5. In vitro translated proteins: A convenient kit for in vitro coupled transcription/translation is commercially available from Promega Corporation. Alternatively, reagents for in vitro transcription and in vitro translation can be purchased separately. For the synthesis of ^{35}S-labeled proteins, follow the instructions provided by the manufacturer. Translated proteins can be stored for up to 2 wk at –70°C.

6. Protein electrophoresis and transfer: A standard protein electrophoresis apparatus and a semi-dry protein transfer apparatus (such as Hoefer™ SemiPhor from Amersham Pharmacia Biotech) are required. Details of protein electrophoresis and transfer protocols can be found in various methods books, such as *Molecular Cloning (17)*.

7. Immunoblotting: Antibodies against many caspase substrates and secondary conjugates are commercially available. The most commonly used caspase substrate is poly(ADP)ribose polymerase (PARP). The anti-PARP antibody supplied by Roche Molecular Biology cleanly detects the 115-kDa PARP precursor and the 89-kDa cleavage product *(18)*. Other common sources of antibodies include Transduction Laboratories, Pharmingen, and Santa Cruz Biotechnology. For the detection of proteins after immunoblotting, enhanced chemiluminescence from Amersham Pharmacia Biotech works well.

3. Methods

3.1. Measurement of Caspase Activity Using Synthetic Peptide Substrates

3.1.1. Preparation of Cell Extracts

For preparation of cell extracts from animal tissue samples, homogenize frozen tissue cut into small pieces in extraction buffer (20 mM piperazine-N,N'-*bis*[2-ethanesulphonic acid], 100 mM NaCl, 10 mM DTT, 1 mM ethylenediamine tetraacetic acid , 0.1% 3-[(3-cholamidopropyl)dimethylammonio]-1-propanesulfonate,

10% sucrose, pH 7.0, containing protease inhibitor cocktail such as, Complete™ from Roche Molecular Biologicals) using a tissue homogenizer prior to cell lysis. For cultured cells in suspension, spin down cells at 200g for 10 min and wash once in ice-cold phosphate-buffered saline (PBS). For adherent cells, gently scrape cells into medium, spin down cell pellet at 200g for 10 min, and wash once in cold PBS. Resuspend cells at approximately 10^7 cells/mL in extraction buffer. Freeze/thaw cells three times and then homogenize by 10 strokes in a glass homogenizer. Spin extracts at 15,000g for 10 min and carefully remove cell extracts leaving the pellet undisturbed. After determining protein concentration the extracts can be frozen at –70°C in small aliquots for several months without any significant loss of caspase activity.

3.1.2. Measurement of Caspase Activity

Caspase assays should be performed continuously if the spectrophotometer is equipped with a regulated temperature chamber that can accommodate cuvettes or 96-well plates; otherwise, the release of AFC, AMC, or pNA can be monitored after a fixed period of incubation. To save reagents, perform assays in a final volume of 50–100 µL. If the fluorometer is equipped with a 96-well plate reader and several assays need to be conducted simultaneously, the reactions can be assembled in a 96-well plate format. If this is not possible, reactions can be conducted in microfuge tubes or cuvettes.

1. Add varying concentrations of the cell lysates to caspase assay buffer supplemented with 0.1 mM of an appropriate caspase substrate and monitor the release or fluorochrome or chromogen at 37°C in the thermostat fitted spectrophotometer (*see* **Notes 1** and **2**). For AMC fluorescence detection, adjust the excitation and emission wavelengths to 385 nm and 460 nm, respectively. For AFC, excitation and emission wavelengths are 400 nm and 505 nm, respectively. pNA absorbance should be monitored at 405–410 nm.
2. Monitor the release of the fluorochrome or chromogen for 30 min. The enzyme activities can be expressed as rate of substrate hydrolysis that can be calculated from the linear portion of the progress curves, prior to the time when substrate depletion slows down the

rate of reaction. If substrate depletion occurs too quickly, dilute cell extracts to get a more linear response.

3. If continuous monitoring of fluorochrome or chromogen release is not possible because of the nonavailability of equipment, assays can be conducted for various lengths of time up to 30 min at 37°C in a waterbath. At the end of the incubation, the reactions can be stopped by adding 0.4 mL of ice-cold water and storing tubes on ice.
4. Plot fluorescence (for AFC or AMC) or absorbance (for pNA) and calculate the rate of hydrolysis from the linear part of the curve.

3.2. Assay of Caspase Activity by Cleavage of ^{35}S Met-Labeled Caspase Substrates

This is a qualitative assay that is suitable for confirming the presence of active caspases in cell extracts. Clone the cDNAs containing caspase cleavage sites, such as PARP *(19,20)*, DNA-PK catalytic subunit *(21)*, or inhibitor of caspase-activated DNase (ICAD) *(22)*, in plasmid vectors that carry either SP6, T3, or T7 promoters (pBluscript, pGEM, and pcDNA3 vectors are all appropriate for this purpose). It is not necessary to clone the entire protein, and truncated coding region containing caspase site(s) that give rise to easily discernible cleavage products work well *(23)*.

1. Purify plasmids using CsCl centrifugation or Qiagen kit and perform in vitro transcription/translation using Promega kit according to the instructions provided by the manufacturer. Typical 50-µL reactions contain 25 µL of TNT lysates, 2 µL of TNT reaction buffer, 1 µL of T3, T7, or SP6 RNA polymerase, 1 µL of amino acid mixture lacking Met, 5 µL of ^{35}S Met, 1 µL of Rnasin, 1 µg plasmid DNA, and sterile RNase-free water.
2. Incubate reaction tubes at 30°C for 1.5–2 h, spin at maximum speed in a microfuge for 5 min, and remove supernatant to fresh tube. In vitro translated proteins can be stored at –70°C for up to 2 wk.
3. For cleavage assays, 5 µL of labeled protein is incubated at 37°C for 2 h with varying amounts of cell extracts (10–50 µg total protein) in caspase assay buffer in a total volume of 20 µL. In control experiments, cell extracts can be preincubated with caspase inhibitors, such as 50 µ*M* zVAD-FMK, for 30 min before the addition of labeled protein substrate.

4. At the end of incubation period, add 20 µL of 2× protein loading buffer (100 mM Tris-HCl pH 6.8, 200 mM dithiothreitol, 20% glycerol, 4% sodium dodecyl sulfate (SDS) 0.2% bromophenol blue) to each tube, boil for 5 min, and centrifuge at maximum speed in a microfuge for 5 min.

5. Remove supernatant to fresh tube and resolve cleavage products by electrophoresis on 10–15% polyacrylamide/SDS gel.

6. After fixation, gels can be dried. Alternatively, proteins can be transferred to polyvinylidine difluoride or nitrocellulose membranes using a semidry transfer apparatus before the visualisation of ^{35}S-labeled protein bands by autoradiography *(24)*. This avoids the possibility of gels cracking during the drying process. In most cases, freshly labeled ^{35}S proteins and their cleavage products can be detected after an overnight exposure to a X-ray film.

3.3. Assessing Caspase-Mediated Substrate Cleavage by Immunoblotting

Because caspase activation results in the cleavage of the caspase precursor into subunits, caspase activation can be indirectly observed by immunoblotting using specific antibodies (*see* **Note 4**). However, a more direct measure of caspase activity, usually that is contributed by the downstream or effector caspases, such as caspase-3 and caspase-7, is to determine whether endogenous caspase targets are being cleaved (*see* **Note 5**). This can be easily achieved by immunoblotting of cell extracts using a specific antibody against a known endogenous caspase substrate. There are close to 300 proteins now known to be cleaved by caspases *(25)*. The most common one, for which good antibodies are available from many commercial suppliers is PARP, a caspase-3 substrate.

1. Prepare samples for electrophoresis by mixing equal volume of protein extract prepared as described in **Subheading 3.1.1.** and 2× protein loading buffer.

2. Cell pellets or small pieces of tissues can also be directly lysed by boiling in 2× protein loading buffer. However, often the lysates prepared in such a way will be very viscous due to the release of DNA. To reduce viscosity, the samples can be passed through a 22-gauge needle three to four times or sonicated for 30 s to shear DNA.

3. Boil samples for 5–10 min and centrifuge lysates for 5 min at maximum speed in a microfuge to remove any insoluble material. At this stage, if required, the samples can be stored at –70°C indefinitely.
4. Electrophorese 10–20 µg of the protein samples on 10% polyacrylamide/SDS gels.
5. Transfer proteins to polyvinylidine difluoride membrane using the semidry protein transfer apparatus.
6. Block membrane in 5% skim milk in PBS containing 0.05% Tween 20 (PBST) for 1 h at room temperature or overnight at 4°C.
7. Dilute primary antibody as suggested by the manufacturer in PBST and incubate the membrane with the antibody solution for 1 h at room temperature.
8. Wash membrane three times for 10 min each and incubate with the appropriate secondary antibody diluted in PBST.
9. For detection of signals by enhanced chemiluminescence, follow instruction supplied by Amersham Pharmacia Biotech. As an example, in healthy cells PARP will appear as a single band of approximately 115 kDa, whereas in cells undergoing apoptosis a gradual decrease in 115-kDa band and appearance of a 89-kDa cleavage product should be clearly visible.

4. Notes

1. Although most caspases are active at pH 7.0, some have different pH optima. For example, caspase-2 and caspase-9 favor a slightly acidic pH *(26)*. If necessary, the assay buffer containing 0.1 *M* MES, pH 6.5, can be used instead of 0.1 *M* 4-(2-hydroxyethyl)-1-piperazineethanesulphonic acid, pH 7.0.
2. If necessary, recombinant caspases expressed in *Escherichia coli* can be used for positive controls. A number of publications describe the preparation of recombinant caspases *(21,24,26)*. Some commercial suppliers, such as Alexis Biochemicals (Switzerland), also provide a number of recombinant purified caspases.
3. To avoid nonspecific hydrolysis of caspase substrates, it is useful to include protease inhibitor cocktail in the cell lysis buffer. Many commercially available protease inhibitor set can be used provided they do not contain caspase inhibitors.
4. To test whether individual caspases are being activated, immunoblot analysis of cell extracts using specific caspase antibodies can be

performed. To do this, prepare cell extract blots as described in **Subheading 3.3.** and probe them with caspase antibodies to determine whether a specific caspase precursor is being cleaved into active subunits. There are numerous commercial sources of caspase antibodies; however, many antibodies on the market are of poor quality. If using a new antibody for the first time, especially when the same antibody has not been used in the published literature, specificity and affinity of the antibody should be empirically established using recombinant caspases. Some antibodies will detect both the precursor and one or more subunits/intermediates, whereas others are specific for either the precursor or the subunits. In some cell types, the half-life of some active caspase subunits is often very short. In such cases a clear decrease in zymogen signal can be seen but not a corresponding increase in the subunit signal.

5. Caspase activation can also be determined *in situ*, in cultured cells, or in tissue sections by immunohistochemistry using antibodies that specifically recognize processed form of caspases. These antibodies are now widely available from various commercial sources.

Acknowledgments

Work in the author's laboratory is supported by the National Health and Medical Research Council of Australia and the Cancer Council of South Australia.

References

1. Kumar, S. (1995) ICE-like proteases in apoptosis. *Trends Biochem. Sci.* **20,** 198–202.
2. Kumar, S. and Lavin, M. F. (1996) The ICE family of cysteine proteases as effectors of cell death. *Cell Death Differ.* **3,** 255–267.
3. Nicholson, D. W. and Thornberry, N. A. (1997) Caspases: killer proteases. *Trends Biochem. Sci.* **22,** 299–306.
4. Nicholson, D. W. (1999) Caspase structure, proteolytic substrates, and function during apoptotic cell death. *Cell Death Differ.* **6,** 1028–1042.
5. Salvesen, G. S. (2002) Caspases and apoptosis. *Essays Biochem.* **38,** 9–19.
6. Zheng, T. S., Hunot, S., Kuida, K., and Flavell, R. A. (1999) Caspase knockouts: matters of life and death. *Cell Death Differ.* **6,** 1043–1053.

7. Wilson, K. P., Black, J-A. F., Thomson, J. A., Kim, E. E., Griffith, J. P., Navia, M. A., et al. (1994) Structure and mechanism of interleukin-1β converting enzyme. *Nature* **370,** 270–275.

8. Walker, N. P. C., Talanian, R. V., Brady, K. D., Dang, L. C., Bump, N. J., Ferenz, C. R., et al. (1994) Crystal structure of the cysteine protease interleukin-1β converting enzyme: a (p20/p10)2 homodimer. *Cell* **78,** 343–352.

9. Rotonda, J., Nicholson, D. W., Fazil, K. M., Gallant, M., Gareau, Y., Labelle, M., et al. (1996) The three-dimensional structure of apopain/CPP32, a key mediator of apoptosis. *Nat. Struct. Biol.* **3,** 619–625.

10. Mittl, P. R., Di Marco, S., Krebs, J. F., Bai, X., Karanewsky, D. S., Priestle, J. P., Tomaselli, K. J., and Grutter, M. G. (1997) Structure of recombinant human CPP32 in complex with the tetrapeptide acetyl-Asp-Val-Ala-Asp fluoromethyl ketone. *J. Biol. Chem.* **272,** 6539–6547.

11. Kumar, S., and Colussi, P. A. (1999) Prodomains-adaptors-oligomerization: the pursuit of caspase activation in apoptosis. *Trends Biochem. Sci.* **24,** 1–4.

12. Kumar, S. (1999) Mechanisms of caspase activation in cell death. *Cell Death Differ.* **6,** 1060–1066.

13. Thornberry, N. A., Rano, T. A., Peterson, E. P., Rasper, D. M., Timkey, T., Garcia-Calvo, M., et al. (1997) A combinatorial approach defines specificities of members of the caspase family and granzyme B. Functional relationships established for key mediators of apoptosis. *J. Biol. Chem.* **272,** 17,907–17,911.

14. Talanian, R. V., Quinlan, C., Trautz, S., Hackett, M. C., Mankovich, J. A., Banach, D., et al. (1997) Substrate specificities of caspase family proteases. *J. Biol. Chem.* **272,** 9677–9682.

15. Dorstyn, L., Colussi, P. A., Quinn, L. M., Richardson, H., and Kumar, S. (1999) DRONC, an ecdysone-inducible Drosophila caspase. *Proc. Natl. Acad. Sci. USA* **96,** 4307–4312.

16. Hawkins, C. J., Yoo, S. J., Peterson, E. P., Wang, S. L., Vernooy, S. Y., and Hay, B. A. (2000) The Drosophila caspase DRONC cleaves following glutamate or aspartate and is regulated by DIAP1, HID, and GRIM. *J. Biol. Chem.* **275,** 27,084–27,093.

17. Sambrook, J., Fritsch, E. F., and Maniatis, T. (1989) *Molecular Cloning: A Laboratory Manual*, 2nd ed. Cold Spring Harbor Laboratory Press, Cold Spring Harbor, New York.

18. Harvey, K. F., Harvey, N. L., Michael, J. M., Parasivam, G., Waterhouse, N., Alnemri, E. S., et al. (1998) Caspase-mediated cleav-

age of the ubiquitin-protein ligase Nedd4 during apoptosis. *J. Biol. Chem.* **273**, 13,524–13,530.

19. Kaufmann, S. H., Desnoyers, S., Ottaviano, Y., Davidson, N. E., and Poirier, G. G. (1993) Specific proteolytic cleavage of poly(ADP-ribose) polymerase: an early marker of chemotherapy-induced apoptosis. *Cancer Res.* **53**, 3976–3985.

20. Lazebnik, Y. A., Kaufmann, S. H., Desnoyers, S., Poirier, G. G., and Earnshaw, W. C. (1994) Cleavage of poly(ADP-ribose) polymerase by a proteinase with properties like ICE. *Nature* **371**, 346–347.

21. Song, Q., Lees-Miller, S. P., Kumar, S., Zhang, N., Chan, D. W., Smith, G. C. M., et al. (1996) DNA-dependent protein kinase catalytic subunit: A target for an ICE-like protease in apoptosis. *EMBO J.* **15**, 3238–3246.

22. Sakahira, H., Enari, M., and Nagata, S. (1998) Cleavage of CAD inhibitor in CAD activation and DNA degradation during apoptosis. *Nature* **391**, 96–99.

23. Harvey, N. L., Butt, A. J., and Kumar, S. (1997) Functional activation of Nedd2/ICH-1 (caspase-2) is an early process in apoptosis. *J. Biol. Chem.* **272**, 13134–13139.

24. Harvey, N. L., Trapani, J. A., Fernandes-Alnemri, T., Litwack, G., Alnemri, E. S., and Kumar, S. (1996) Processing of the Nedd2 precursor by ICE-like proteases and granzyme B. *Genes Cells* **1**, 673–685.

25. Fischer, U., Janicke, R. U., and Schulze-Osthoff, K. (2003) Many cuts to ruin: a comprehensive update of caspase substrates. *Cell Death Differ.* **10**, 76–100.

26. Garcia-Calvo, M., Peterson, E. P., Rasper, D.M., Vaillancourt, J. P., Zamboni, R., Nicholson, D. W., et al. (1999) Purification and catalytic properties of human caspase family members. *Cell Death Differ.* **6**, 362–369.

3

Flow Cytometry-Based Methods for Apoptosis Detection in Lymphoid Cells

Owen Williams

Summary

Apoptosis is an active form of cell death that plays a critical role in lymphocyte development, selection and homeostasis. This process is characterized by the activation of biochemical pathways that lead to changes in cellular morphology (including cell shrinkage, membrane blebbing and nuclear condensation), DNA fragmentation, perturbation of mitochondrial membrane function and changes in the plasma membrane. Each of these cellular alterations can be rapidly quantitated in lymphocyte apoptosis using flow cytometry.

Key Words: Lymphocytes; cell death; FACS; DNA fragmentation; mitochondria.

1. Introduction

Apoptosis is an active form of cell death that is of fundamental importance to the biology of various cellular systems (*1*). This process plays a critical role in lymphocyte development, selection, and homeostasis (*2–4*). Apoptosis is characterized by the activation of bicohemical pathways that lead to changes in cellular morphology (including cell shrinkage, membrane blebbing, and nuclear condensation), DNA fragmentation, perturbation of mitochondrial membrane function, and changes in the plasma membrane. The apoptotic bodies produced by this form of cell

From: *Methods in Molecular Biology, vol. 282: Apoptosis Methods and Protocols*
Edited by: H. J. M. Brady © Humana Press Inc., Totowa, NJ

death are rapidly cleared in vivo but may persist in vitro for some time.

Each of these cellular alterations can be rapidly quantitated in lymphocyte apoptosis using flow cytometry. Changes in cellular morphology can be detected by measuring alterations in the way apoptotic cells scatter the light of the laser beam in a flow cytometer (5). Cell shrinkage results in a decrease in forward light scatter (FSC) and nuclear condensation causes a transient increase in right angle or side scatter (SSC), followed by reduced SSC during the final stages of apoptosis. Alterations in plasma membrane permeability can be measured by incorporation of 7-aminoactinomycin D (7AAD; ref. 6). Analysis of light scatter can be combined with 7AAD incorporation in a simple and accurate technique for detecting apoptotic lymphocytes (7,8).

Reduced staining of apoptotic cells with DNA intercalating fluorochromes, such as propidium iodide (PI) or 7AAD, can be used to follow DNA fragmentation (9–11). In these assays, diminished DNA staining of apoptotic cells gives rise to a hypodiploid (sub-G_0/G_1) peak, most probably caused by the loss of oligo-nucleosomal fragments from dying cells during the permeabilization and washing steps of these methods (12).

Changes in mitochondrial membrane function during apoptosis affect the inner membrane and result in a loss of mitochondrial transmembrane potential. This can be measured by incubation of cells with the potential-sensitive dye 3,3'-dihexyloxacarbocyanide iodide [$DiOc_6(3)$; 13,14]. Mitochondrial interaction with this dye is dependent on the maintenance of high transmembrane potential and its uptake is reduced in apoptotic lymphocytes (15).

Plasma membrane phospholipid asymmetry is lost in cells undergoing apoptosis, leading to the exposure of phosphatidylserine (PS) to the outer leaflet of the membrane. Exposure of PS on the plasma membrane can be detected using the PS-binding anticoagulant annexin V labeled with fluorescein isothiocyanate (FITC) or phycoerythrin (PE; refs. 16,17).

2. Materials

2.1. Instrumentation and General Buffers for Flow Cytometry

1. Flow cytometer equipped with a single 488-nm argon laser, and 530-nm, 585-nm, and >650-nm filters for analysis in FL1, FL2, and FL3 channels, respectively.
2. Phosphate-buffered saline (PBS): Prepare in 1-L quantities and store at room temperature: 140 mM NaCl, 2.7 mM KCl, 8 mM Na$_2$HPO$_4$, 1.5 mM KH$_2$PO$_4$.
3. Staining buffer: Make up in 100-mL quantities, store at 4°C: PBS, 1% w/v bovine serum albumin (Sigma-Aldrich Chemical Co.), 0.1% w/v NaN$_3$. Note: NaN$_3$ is toxic.

2.2. 7AAD Stock

1. 1 mg/mL 7AAD stock: 7AAD is from Sigma-Aldrich Chemical Co. Make up in 50% v/v DMSO (analytical grade) in 1-mL quantities. Store at 4°C in the dark. Note: 7AAD is toxic.

2.3. Buffers for DNA Staining With PI

1. 1 mg/mL PI stock: Make in H$_2$O in 10-mL quantities. Store at 4°C in the dark. Note: PI is toxic.
2. Hypotonic PI buffer: Prepare in 500-mL quantities and store at 4°C: 50 µg/mL PI, 0.1% w/v sodium citrate, 0.1% w/v Triton X-100 (Sigma-Aldrich Chemical Co.).

2.4. Buffers for DNA Staining With 7AAD

1. Modified staining buffer: Prepare in 100-mL quantities, store at 4°C: PBS, 2% v/v heat-inactivated fetal calf serum (Life Technologies), 0.1% w/v NaN$_3$.
2. 7AAD saponin buffer: Prepare in 100-mL quantities, store at 4°C: add 0.3% w/v saponin (Sigma-Aldrich Chemical Co.) and 4 µg/mL 7AAD to the modified staining buffer.

2.5. DiOc$_6$ Stock

1. 4 µM DiOc$_6$(3) stock: DiOc$_6$(3) is from Molecular Probes. Prepare in ethanol (analytical grade) in 1-mL quantities. Store at 4°C in the dark.

2.6. Buffer for Annexin V Assay

1. FITC or PE-labeled annexin V can be obtained from Pharmingen.
2. Binding buffer: Make up in 100-mL quantities, store at 4°C: 10 mM
   N-hydroxyethylpiperazine-N'-2-ethanesulfonate, pH 7.4; 150 mM
   NaCl; 5 mM KCl; 1 mM MgCl$_2$; 1.8 mM CaCl$_2$.

3. Methods

3.1. Light Scatter Analysis

1. Resuspend 1×10^6 cells in 1 mL of PBS, 0.1% w/v NaN3. Analyze
   FSC vs SSC, both on linear scales, of cells using a flow cytometer.
2. Set a gate for viable cells on the FSC vs SSC dot plot, by analyzing
   control viable cells (region 1, **Fig. 1A**). Cells in early stages of
   apoptosis will show a decrease in FSC and an increase in SSC
   (region 2, **Fig. 1B**) compared with viable cells. Dead cells and late-
   stage apoptotic cells will show a decrease in both FSC and SSC
   (region 3, **Fig. 1B**).

3.2. Light Scatter Analysis Combined With 7AAD Uptake

1. Resuspend 1×10^6 cells in 1 mL of PBS, 0.1% w/v NaN$_3$.
2. Add 20 μL of 7AAD stock (final 7AAD concentration is 20 μg/mL)
   and incubate cells for 20 min at 4°C in the dark.
3. Analyze cells using a flow cytometer and by detecting 7AAD emission (655 nm) in the FL3 channel on a logarithmic scale.
4. Use control viable cells to set a viable gate (region 1, **Fig. 2A**). Viable
   cells are negative for 7AAD staining. Cells in early stages of
   apoptosis will show a decrease in FSC and intermediate staining with
   7AAD (region 2, **Fig. 2B**). Late-stage apoptotic cells and dead cells
   will have low FSC and high 7AAD staining (region 3, **Fig. 2B**). Cell
   fragments and apoptotic bodies will have very low FSC and intermediate 7AAD (region 4, **Fig. 2B**).

3.3. DNA Staining With PI

1. Aliquot 1×10^6 cells into 1.5-mL Eppendorf tubes, centrifuge (200g),
   and resuspend the cell pellet in 0.5 mL of cold PBS.
2. Repeat the centrifugation and resuspend the cells in 0.5 mL of hypotonic PI buffer by gently inverting the tube.
3. Incubate overnight at 4°C in the dark.

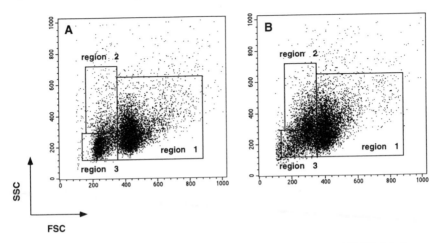

Fig. 1. Light scatter analysis of apoptotic lymphocytes. (**A**) Viable lymphocytes; (**B**) apoptotic lymphocytes.

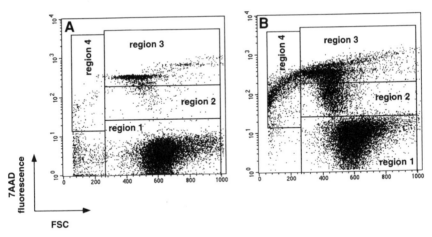

Fig 2. Light scatter/7AAD analysis. (**A**) Viable lymphocytes; (**B**) apoptotic lymphocytes.

4. Analyze stained nuclei with a flow cytometer detecting PI emission (625 nm) in the FL2 channel on a logarithmic scale.
5. Set the PI fluorescence peak corresponding to nuclei from resting viable (G_0/G_1) cells at channel 100 (region 1, **Fig. 3A**). Exclude cell debris by increasing the FSC threshold until events in 1–100 channels are <1%, using viable cells. Nuclei from apoptotic cells will give a

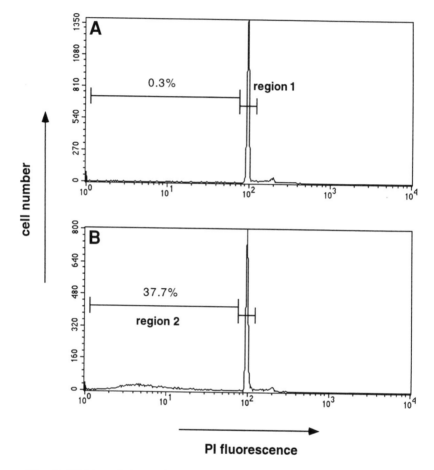

Fig. 3. DNA staining with PI. (**A**) Viable lymphocytes; (**B**) apoptotic lymphocytes. Numbers refer to the percentage of apoptotic cells in each sample.

(subdiploid) PI fluorescence peak (region 2, **Fig. 3B**) below the G_0/G_1 peak (*see* **Note 1**).

3.4. DNA Staining With 7AAD
Combined With Cell Surface Staining

1. Aliquot 5×10^5 cells into wells of a V-bottomed 96-well plate. Surface stain cells with appropriate antibodies in modified staining buffer.

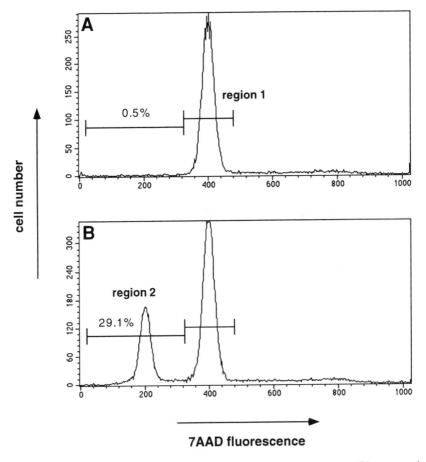

Fig. 4. DNA staining with 7AAD. (**A**) viable lymphocytes; (**B**) apoptotic lymphocytes. Numbers refer to the percentage of apoptotic cells in each sample.

2. Wash cells (centrifugation at 200*g*) once with PBS, 0.1% w/v NaN$_3$ and resuspend in 100 µL of 7AAD saponin buffer (*see* **Note 2**).
3. Incubate for 30 min at 37°C in the dark. Cells can be analyzed immediately, without washing, or stored at 4°C overnight in the dark.
4. Analyze cells with a flow cytometer detecting 7AAD emission in the FL3 channel on a linear scale. Antibodies used in surface staining are detected in the FL1 and FL2 channels.

5. Set the 7AAD peak corresponding to resting viable (G_0/G_1) cells at channel 400 of a 1024 channel scale (region 1, **Fig. 4A**). Apoptotic cells will give a (subdiploid) 7AAD fluorescence peak (region 2, **Fig. 4B**) below the G_0/G_1 peak (*see* **Notes 1** and **3**).

3.5. Analysis of Mitochondrial Transmembrane Potential Using DiOc$_6$(3)

1. Resuspend 5×10^6 cells in 1 mL of the appropriate culture medium and add 10 µL of DiOc$_6$(3) stock [final DiOc$_6$(3) concentration is 40 nm].
2. Incubate cells for 15 min at 37°C in the dark.
3. Analyze cells immediately with a flow cytometer, detecting DiOc$_6$(3) emission in the FL2 channel on a logarithmic scale.
4. Set the DiOc$_6$(3) peak of viable cells between channels 100–1000 (region 1, **Fig. 5A**). Apoptotic cells, with decreased mitochondrial transmembrane potential, show reduced DiOc$_6$(3) fluorescence (region 2, **Fig. 5B**; *see* **Note 4**).

3.6. Analysis of PS Exposure Using Annexin V

1. Aliquot 1×10^5 cells into wells of a V-bottomed 96-well plate. Surface stain cells with appropriate antibodies in staining buffer.
2. Wash cells (centrifugation at 200g) once with PBS, 0.1% w/v NaN$_3$ and resuspend in 200 µL of binding buffer.
3. Add annexin V (FITC or PE labeled) to a final concentration of 1 µg/mL.
4. Incubate cells for 15 min at 37°C in the dark.
5. Analyze cells immediately, or within 1 h of staining, with a flow cytometer detecting annexin V staining in the appropriate FL channel (FL1 for FITC and FL2 for PE) on a logarithmic scale.
6. Set the annexin V-FITC, or -PE, peak of viable cells between channel 1–100 (region 1, **Fig. 6A**). Apoptotic cells will show increased binding of annexin V and increased annexin V-FITC, or -PE, fluorescence (region 2, **Fig. 6B**).

4. Notes

1. The G_0/G_1 DNA fluorescence peak of cells stained with PI (**Subheading 3.3.**) or 7AAD (**Subheading 3.4.**) can sometimes shift in intensity between samples. This is a particular problem when com-

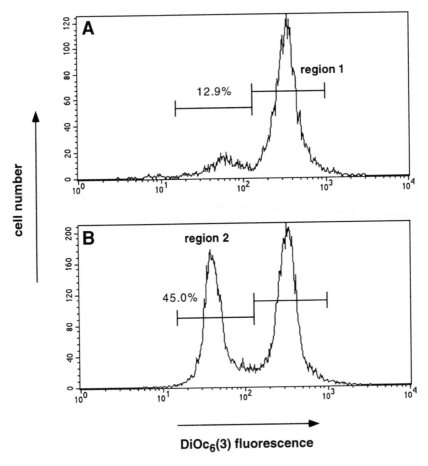

Fig. 5. Analysis of mitochondrial transmembrane potential with DiOc$_6$(3). (**A**) Viable lymphocytes; (**B**) apoptotic lymphocytes. Numbers refer to the percentage of apoptotic/dead cells in each sample. *See* **Notes** for how to distinguish between cells in early stages of apoptosis and necrotic/late-stage apoptotic cells.

paring samples where the majority of cells have undergone apoptosis with control viable cells. Reducing the number of cells stained and/ or increasing the volume in which they are stained will reduce this effect.

2. Some batches of saponin used in cell permeabilization, when staining DNA with 7AAD (**Subheading 3.4.**), can affect cell surface antibody staining. If samples treated with saponin show a loss of cell

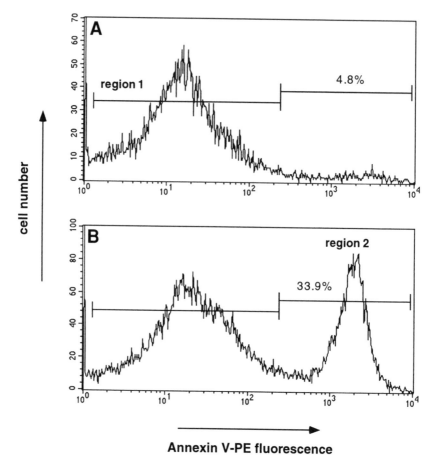

Fig. 6. Analysis of PS exposure with annexin V-PE. (**A**) Viable lymphocytes; (**B**) apoptotic lymphocytes. Numbers refer to the percentage of
apoptotic/dead cells in each sample. *See* **Note 4** for how to distinguish between
cells in early stages of apoptosis and necrotic/late-stage apoptotic cells.

surface staining when compared with nonpermeabilized stained cells,
the saponin concentration in the 7AAD saponin buffer should be
reduced (in the 0.3–0.03% range).

3. The subdiploid DNA fluorescence peak of apoptotic cells stained
with 7AAD (**Subheading 3.4.**) can sometimes merge with the y-axis,
with many apoptotic events thus being lost from the analysis. This
can be avoided by detecting 7AAD emission on a logarithmic scale,
in the FL3 channel.

4. Staining with DiOc$_6$(3) (**Subheading 3.5.**) and annexin V (in **Subheading 3.6.**) may be combined with PI staining to distinguish cells in early stages of apoptosis from cells undergoing secondary necrosis and late-stage apoptotic cells. Add PI to samples, to a final concentration of 5 μg/mL, in the last 5 min of incubation at 37°C with DiOc$_6$(3) or annexin V. PI emission is detected in the FL2 when used in combination with annexin V-FITC and in the FL3 channel when used in combination with annexin V-PE or DiOc$_6$(3). Viable cells will be negative for PI and negative for annexin V or positive for DiOc$_6$(3). Cells in early stages of apoptosis will be negative for PI but will show increased annexin V or decreased DiOc$_6$(3) fluorescence. Dead cells and late-stage apoptotic cells will be positive for PI, and positive for annexin V or negative for DiOc$_6$(3).

References

1. Wyllie, A. H., Kerr, J. F., and Currie, A. R. (1980) Cell death: the significance of apoptosis. *Int. Rev. Cytol.* **68,** 251–306.
2. Smith, C. A., Williams, G. T., Kingston, R., Jenkinson, E. J., and Owen, J. J. (1989) Antibodies to CD3/T-cell receptor complex induce death by apoptosis in immature T cells in thymic cultures. *Nature* **337,** 181–184.
3. Liu, Y. J., Joshua, D. E., Williams, G. T., Smith, C. A., Gordon, J., and MacLennan, I. C. (1989) Mechanism of antigen-driven selection in germinal centres. *Nature* **342,** 929–931.
4. MacDonald, H. R. and Lees, R. K. (1990) Programmed death of autoreactive thymocytes. *Nature* **343,** 642–644.
5. Swat, W., Ignatowicz, L., and Kisielow, P. (1991) Detection of apoptosis of immature CD4+8+ thymocytes by flow cytometry. *J. Immunol. Methods* **137,** 79–87.
6. Schmid, I., Krall, W. J., Uittenbogaart, C. H., Braun, J., and Giorgi, J. V. (1992) Dead cell discrimination with 7-amino-actinomycin D in combination with dual color immunofluorescence in single laser flow cytometry. *Cytometry* **13,** 204–208.
7. Philpott, N. J., Turner, A. J., Scopes, J., Westby, M., Marsh, J. C., Gordon-Smith, E. C., et al. (1996) The use of 7-amino actinomycin D in identifying apoptosis: simplicity of use and broad spectrum of application compared with other techniques. *Blood* **87,** 2244–2251.

8. Lecoeur, H., Ledru, E., Prevost, M. C., and Gougeon, M. L. (1997) Strategies for phenotyping apoptotic peripheral human lymphocytes comparing ISNT, annexin-V and 7-AAD cytofluorimetric staining methods. *J. Immunol. Methods* **209,** 111–123.

9. Nicoletti, I., Migliorati, G., Pagliacci, M. C., Grignani, F., and Riccardi, C. (1991) A rapid and simple method for measuring thymocyte apoptosis by propidium iodide staining and flow cytometry. *J. Immunol. Methods* **139,** 271–279.

10. Telford, W. G., King, L. E., and Fraker, P. J. (1992) Comparative evaluation of several DNA binding dyes in the detection of apoptosis-associated chromatin degradation by flow cytometry. *Cytometry* **13,** 137–143.

11. Zal, T., Volkmann, A., and Stockinger, B. (1994) Mechanisms of tolerance induction in major histocompatibility complex class II-restricted T cells specific for a blood-borne self-antigen. *J. Exp. Med.* **180,** 2089–2099.

12. Darzynkiewicz, Z., Bruno, S., Del Bino, G., Gorczyca, W., Hotz, M. A., Lassota, P., and Traganos, F. (1992) Features of apoptotic cells measured by flow cytometry. *Cytometry* **13,** 795–808.

13. Johnson, L. V., Walsh, M. L., Bockus, B. J., and Chen, L. B. (1981) Monitoring of relative mitochondrial membrane potential in living cells by fluorescence microscopy. *J. Cell Biol.* **88,** 526–535.

14. McGinnes, K., Chapman, G., and Penny, R. (1988) Effects of interferon on natural killer (NK) cells assessed by fluorescent probes and flow cytometry. *J. Immunol. Methods* **107,** 129–136.

15. Zamzami, N., Marchetti, P., Castedo, M., Zanin, C., Vayssiere, J. L., Petit, P. X., and Kroemer, G. (1995) Reduction in mitochondrial potential constitutes an early irreversible step of programmed lymphocyte death in vivo. *J. Exp. Med.* **181,** 1661–1672.

16. Koopman, G., Reutelingsperger, C. P., Kuijten, G. A., Keehnen, R. M., Pals, S. T., and van Oers, M. H. (1994) Annexin V for flow cytometric detection of phosphatidylserine expression on B cells undergoing apoptosis. *Blood* **84,** 1415–1420.

17. Vermes, I., Haanen, C., Steffens-Nakken, H., and Reutelingsperger, C. (1995) A novel assay for apoptosis. Flow cytometric detection of phosphatidylserine expression on early apoptotic cells using fluorescein labelled Annexin V. *J. Immunol. Methods* **184,** 39–51.

4

Methods of Analyzing Chromatin Changes Accompanying Apoptosis of Target Cells in Killer Cell Assays

Richard C. Duke

Summary

Natural killer (NK) cells are a crucial component of the immune system. For example, the NK cell-mediated killing of tumor cells is a first line of defense against the development of cancer. Detection and quantification of apoptosis induced in target cells by NK cells is a useful tool for studies of the mechanisms of both immunological protection and pathogenesis. This chapter describes how to do this by using techniques with a broad application to both NK cells and cytotoxic T cells.

Key Words: Natural killer cells; target cells; effector cells; perforin; Fas; cytolysis.

1. Introduction

Natural killer (NK) cells kill their targets by inducing two distinct modes of cell death—apoptosis and necrosis *(1–4)*. The end result in both cases is target cell lysis, broadly defined as a loss of membrane integrity and most usually assessed by either chromium release or uptake of certain nonvital dyes such as eosin, trypan blue, or propidium iodide.

Apoptosis occurs under normal physiological conditions and requires the active participation of the cell itself—death occurs from within (reviewed in **ref. 5**; *see* also **Notes 1** and **2**). Apoptotic cell

From: *Methods in Molecular Biology, vol. 282: Apoptosis Methods and Protocols*
Edited by: H. J. M. Brady © Humana Press Inc., Totowa, NJ

death can be induced by a variety of stimuli; however, NK cells are generally thought to use two mechanisms (*see* **Note 3** for further reference). First, NK cells can induce apoptosis through secretion of granzymes acting in concert with perforin *(3,6)*. Second, NK cells can express a ligand for Fas (also called CD95 or Apo-1) allowing them to induce apoptosis in target cells that express the functional form of the receptor *(7,8)*. Induction of apoptosis by both mechanisms involves activation of a family of proteases in the target cell termed caspases (*see* also Chapter 14). Caspases cleave a variety of cellular targets including nuclear and cytoskeletal components. The end result is that the cell undergoing apoptosis shows a progressive contraction of cell volume often accompanied by violent blebbing or zeiosis *(5)*. The structural integrity of most cytoplasmic organelles is often preserved but dramatic changes occur in the nucleus. These changes involve condensation and fragmentation of chromatin into uniformly dense masses that can be readily detected microscopically using DNA-binding dyes (**Subheading 3.1.**; *see* **Note 2** for further information regarding the importance of morphological assays). In many cells undergoing apoptosis, extensive double-stranded DNA cleavage also occurs although there are exceptions to this rule (*see* **Note 4**). The assays described in **Subheadings 3.2., 3.3.**, and **3.4.** take advantage of the fact that chromatin composed of fragmented DNA can be physically separated from intact chromatin. The TUNEL assay, described in **Subheading 3.5.**, detects double-stranded DNA cleavage at the single-cell level by using terminal deoxynucleotidal transferase (TdT) to add fluorescein isothiocyanate (FITC)-labeled dUTP at the cleavage site; positive cells are detected by microscopy or by flow cytometry. It is important to understand that a cell that has undergone apoptosis will eventually lyse. Temporally, therefore, one observes changes in nuclear structure that precede loss of membrane integrity; e.g., DNA fragmentation is detected prior to chromium release and chromatin condensation is observed microscopically before nonvital dyes are taken up *(9,10)*.

In in vitro assays, it is easiest to consider that necrosis occurs when a cell lyses in the absence of demonstrable apoptotic changes.

Necrosis results when a cell is unable to maintain homeostasis, leading to an influx of extracellular ions and water. Intracellular organelles, most notably the mitochondria, and the entire cell swell and eventually rupture. In high concentrations, purified perforin induces necrosis but not apoptosis *(11)*. It has been shown that NK cells can induce cytolysis in the absence of apoptotic changes, strongly suggesting that perforin is the causative agent *(1,12; see* **Note 3** for further information). NK cells, as compared to cytotoxic T lymphocytes (CTLs), tend to be more likely to induce purely necrotic cell death because they express relatively higher amounts of perforin than do CTLs *(13)*.

Because the mechanism of apoptosis is poorly understood at the present time, it is probably best to perform several assays to confirm an observation of apoptotic cell death (*see* **Note 2**). In addition, it is important to employ assays that detect cytolysis concomitantly with apoptosis assays to confirm that NK cell-mediated killing has occurred.

2. Materials

1. Tissue culture medium (TCM): In all instances, TCM refers to the culture medium which is typically employed in an investigator's laboratory to grow tumor cells and to culture primary lymphocytes.
2. Dye mixes for quantification of apoptotic cell death using fluorescent dyes and UV microscopy (**Subheading 3.1.**) (*see* **Note 5** regarding handling of these chemicals):
   a. Individual stock solutions of acridine orange (AO; Sigma Chemical, St. Louis, MO); ethidium bromide (EB, Sigma); propidium iodide (PI, Sigma); or Hoechst 33342 (Hoe; Sigma) are prepared in deionized H_2O at a concentration of 1 mg/mL. Store stock solutions at 4°C for up to 12 mo and protected from light.
   b. For use in assays, prepare working solutions of 100 µg/mL of AO + 100 µg/mL of EB (AO+EB) or 100 µg/mL of Hoe + 100 µg/mL of PI (Hoe + PI) in deionized H_2O. Working solutions can be used for up to 6 mo if stored at 4°C and protected from light.

3. Radiolabeled compounds for quantifying DNA fragmentation and lysis (**Subheadings 3.2.** and **3.3.**):

   a. 1 mCi/mL of iododeoxyuridine (^{125}IUdR) in aqueous solution (2000 Ci/mmol; ICN Pharmaceuticals, Costa Mesa, CA).

   b. 1 mCi/mL of [methyl-^{3}H]thymidine ([^{3}H]TdR) in aqueous solution (2.0 Ci/mmol; ICN).

   c. 5 mCi/mL of ^{51}Cr as Na_2CrO_4 in saline (200–900 Ci/g; ICN).

4. TTE solution for quantifying DNA fragmentation and lysis (**Subheading 3.2.**): 10 mM Tris HCl, 1 mM EDTA; 0.2% Triton X-100, pH 7.4 (store at 4°C).

5. Buffers and reagents for agarose gel electrophoresis (**Subheding 3.4.**):

   a. TE buffer: 10 mM Tris HCl, 1 mM EDTA, pH 7.4 (store at 4°C).

   b. TTE solution: TE buffer containing 0.2% Triton X-100 (store at 4°C).

   c. 5 mM NaCl (store at 4°C).

   d. Isopropanol (store at –20°C).

   e. 70% Ethanol (store at –20°C).

   f. 10X DNA loading buffer (store at RT): 20% Ficoll 400, 0.1 M EDTA, 1% sodium dodecylsulfate (SDS), 0.25% bromphenol blue, and 0.25% xylene cyanol FF.

   g. 10X TBE electrophoresis buffer stock solution (1 L): 108 g of Tris base, 55 g of boric acid, and 100 mL of 0.2 M EDTA, pH 8.0. Store at room temperature.

6. Buffers and reagents for TUNEL assay (**Subheading 3.5.**) (*see* **Note 6** regarding use of these reagents):

   a. TdT reaction buffer: 0.5 M cacodylic acid, sodium salt, pH 6.8, 1 mM CoCl$_2$, 0.5 mM dithiothreitol (DTT), 0.05% (w/v) bovine serum albumin (BSA), and 0.15 M NaCl. Store at room temperature or 4°C for several months.

   b. TdT/FITC–dUTP reaction mixture: TdT reaction buffer containing 1 μM FITC–dUTP (fluorescein-12-dUTP, Roche Molecular Biochemicals, Indianapolis, IN) and 2–4 U TdT (terminal deoxynucleotidal transferase, Roche Molecular Biochemicals), prepared immediately before use.

   c. 4% Paraformaldehyde prepared in PBS.

7. Reagents for detecting Fas-mediated apoptosis (**Subheading 3.6.**):

   a. EGTA/MgCl$_2$ stock solution: 500 mM EGTA (tetrasodium EGTA, Sigma), 1 M MgCl$_2$, pH 7.0–7.5 (store at 4°C).

    b. Soluble human Fas–human F$_c$ (Fas–F$_c$, Immunex, Seattle, WA) prepared at 1 mg/mL in ddH$_2$O (store at –20°C; avoid subjecting to multiple rounds of freezing and thawing).

    c. Anti-human Fas ligand antibody (Clone 4H9, Immunotech, Westbrook, ME) prepared at 1 mg/mL in ddH$_2$O (store at –20°C; avoid subjecting to multiple rounds of freezing and thawing).

    d. PMA stock solution (1000X): 5 µg/mL PMA (phorbol 12-myristate 13-acetate, Sigma) dissolved in dimethyl sulfoxide (DMSO) (store at –20°C; avoid subjecting to multiple rounds of freezing and thawing).

    e. Ionomycin stock solution (1000X): 500 µg/mL of Ionomycin (Sigma) dissolved in DMSO (store at –20°C; avoid subjecting to multiple rounds of freezing and thawing).

3. Methods

3.1. Quantification of Apoptotic Cell Death Using Fluorescent Dyes and UV Microscopy

In this protocol, fluorescent DNA-binding dyes are added to a mixture of effector and target cells and examined by fluorescence microscopy to visualize and count cells with aberrant chromatin distribution. Acridine orange and Hoechst 33342 are used to quantify how many cells have undergone apoptosis, but they cannot differentiate between intact (viable) and lysed (nonviable) cells. To do the latter, a mixture of acridine orange and ethidium bromide or Hoechst 33342 and propidium iodide is used.

Both live and dead cells take up acridine orange and Hoechst 33342, whereas only dead cells take up ethidium bromide and propidium iodide. AO intercalates into DNA, making it appear bright green, and binds to RNA in the cytoplasm, staining it red. Hoe binds only to DNA, making it appear bright blue. EB and PI intercalate into DNA, making it appear orange, but bind only weakly to RNA, which may appear slightly red. Thus a viable cell stained with AO + EB will have bright green chromatin and red cytoplasm, whereas a dead cell will have bright orange chromatin (EB overwhelms AO) and its cytoplasm, if it has any RNA remaining, will appear dark red. Similarly, a viable cell stained with Hoe + PI will have blue chromatin and no cytoplasmic staining, whereas a dead cell will have bright pink chromatin (Hoe and PI staining blend) and its cytoplasm, if it has any RNA remaining, will appear dark red.

It is extremely important to acquaint oneself with normal chromatin distribution in the target cell line being investigated. When stained with DNA-binding dyes, both live and dead nonapoptotic (normal) cell nuclei will have "structure;" variations in fluorescence intensity reflect the distribution of euchromatin and heterochromatin (*see* **Note 2**). Apoptotic nuclei, in contrast, have highly condensed chromatin that is uniformly fluorescent. This can take the form of crescents around the periphery of the nucleus, or the entire chromatin can be present as one or a group of featureless, bright spherical beads. In advanced apoptosis, the cell will have lost DNA or become fragmented into "apoptotic bodies" and overall brightness will be less than that of a normal cell.

This method can also be used to evaluate factors released by NK cells or isolated from lytic granules by incubating target cells with the material of interest prior to assessment of apoptosis.

Detection of Hoechst 33342 requires UV excitation wavelengths of less than 350 nm which may not be available on standard fluorescent microscopes.

3.1.1. Effector and Target Cells

In this protocol, it is important to be able to distinguish between effector and target cells by nuclear morphology. This is not as foreboding as might first be thought. Tumor cells are often used as targets in NK assays, and these have distinct nuclear morphologies compared to lymphocyte effector cell populations. In general, tumor cells, even those of hematopoietic origin, have much larger nuclei than normal lymphocytes and can be easily differentiated on this basis. Nonetheless, it is best to use highly purified effector cell populations and/or clones (*see* Chapter 1–5 and **ref. 8–10**) such that effector:target (E:T) ratios of less than 10:1, or preferably 1:1, are employed when utilizing this protocol.

Although this assay may appear too cumbersome for routine assays, it should not be overlooked as it provides a wealth of information. Target and effector cells can be coincubated for several days as there is no need to worry about spontaneous release of radioactive labels. This assay also provides information regarding the effect of targets on effector cell populations.

3.1.2. Assay

1. Using a minimum of 5×10^4 target cells per condition, incubate effector and target cells together as appropriate in TCM. Include the following controls: target cells without effectors and effector cells without targets. Targets and effectors can be incubated together in 96-well plates or tissue culture tubes.
2. At the end of the assay period, pellet cell mixtures by centrifugation ($200g$; 10 min) and carefully remove all but approx 25–30 μL of the supernatant.
3. Resuspend the cell pellet and transfer cell suspension to a 12×75-mm glass tube containing 2 μL of either AO + EB or Hoe + PI.
4. Place 10 μL of this mixture on a clean microscope slide and coverslip. Examine with a 40× to 60× objective using epiilumination and a filter combination suitable for observing fluorescein (in the case of AO + EB) or Dapi (in the case of Hoe + PI).
5. Count a minimum of 200 total target cells and record the number of each of the four cellular states: (a) live target cells with normal nuclei (LN: bright green [AO + EB] or blue [Hoe + PI] chromatin with organized structure); (b) live cells with apoptotic nuclei (LA: bright green [AO + EB] or blue [Hoe + PI] chromatin that is highly condensed or fragmented); (c) dead cells with normal nuclei (DN; bright orange [AO + EB] or pink [Hoe + PI] chromatin with organized structure); and (d) dead cells with apoptotic nuclei (DA; bright orange [AO + EB] or pink [Hoe + PI] chromatin that is highly condensed or fragmented).
6. Determine apoptotic index and percentage of necrotic cells using the following formula:

$$\% \text{ apoptotic cells (apoptotic index)} = \frac{LA + DA}{LN + LA + DN + DA} \times 100$$

$$\% \text{ necrotic cells} = \frac{DN}{LN + LA + DN + DA} \times 100$$

$$\% \text{ dead cells} = \frac{DN + DA}{LN + LA + DN + DA} \times 100$$

3.2. Quantification of DNA Fragmentation and Lysis Using Cells Containing Radiolabeled DNA

The following protocol takes advantage of the size difference between fragmented DNA and intact chromatin to quantify target cell apoptosis. Target cell DNA is discriminated from effector cell DNA by metabolically labeling the target cell DNA with ^{125}IUdR prior to performing the assay.

In this protocol, fragmented DNA released from the nuclei of lysed cells is separated from intact chromatin by centrifugation. In brief, target and effector cells are incubated together for various times and the quantification assay is then performed in two steps. In the first step, intact target cells and large cell debris are pelleted by centrifugation. The supernatant, termed "S," is carefully withdrawn and saved. The "S" fraction contains fragmented DNA and cytoplasmic contents (including chromium) released from target cells that lysed during the assay period. Because apoptotic cells may not have lysed during the assay period, the cell pellet is treated with a hypotonic buffer containing EDTA and Triton which results in lysis of the cells and release of their nuclei. The lysate is then subjected to centrifugation, which results in sedimentation of intact nuclei and chromatin. The supernatant from the lysates, termed "T" for "top," contains fragmented DNA released from the nuclei of apoptotic cells as well as the remaining chromium. The pellet, termed "B" for "bottom," contains predominantly intact, chromosome-length chromatin. It should be noted that separation is not absolute as some fragmented DNA can be found in the intact fraction.

By utilizing cells which are doubly labeled with ^{125}IUdR and ^{51}Cr, simultaneous measurement of DNA fragmentation and lysis in the same target cell populations is possible.

3.2.1. DNA and Cytoplasmic Radiolabeling With ^{125}I-UdR and ^{51}Cr

1. Subculture target cells to be labeled at $1–4 \times 10^5$ cells/mL in TCM the day before the assay is to be performed. It is necessary that the cells be growing well (*see* **Note 7** for further information regarding DNA labeling).

2. Transfer cells to 17×100-mm polystyrene (Falcon) or 15-mL coni-cal polypropylene tissue culture tubes and pellet by centrifugation ($200g$; 10 min). Use polypropylene tubes for target cells that are nor-mally adherent.

3. Aspirate supernatant and resuspend in 200 µL TCM, so that concen-tration is $1–4 \times 10^6$ cells/mL.

4. Add 2 to 20 µCi (2–20 µL) of ^{125}I-UdR and incubate cells 90–120 min at 37°C. Add 100 µCi of ^{51}Cr during the last 60 min of the incu-bation period.

5. Wash cells 2× with 10 mL of TCM (prewarmed to 37°C). Resuspend cells in 10 mL of TCM and incubate for 30 min at 37°C. Pellet cells and resuspend in TCM at 1×10^5 cells/mL for assay.

6. Determine level of incorporation by measuring radioactivity in 10,000 target cells with a gamma counter. This procedure should yield an incorporation level of 0.5–3 cpm ^{125}I/cell and 0.05–0.5 cpm ^{51}Cr/cell. If lower levels of ^{125}I-UdR incorporation are observed, the cells may be infected with mycoplasma or may have not been sub-cultured at a low enough initial density. Use doubly labeled cells for experimentation only if the ratio of ^{125}I to ^{51}Cr is greater than 2:1.

7. *Note:* This assay was developed using a Beckman Gamma 5000 counter equipped with a ^{51}Cr Iso-Set Module and a Variable Dis-criminating Module. The ^{51}Cr module detects only ^{51}Cr and not ^{125}I. The variable module was calibrated so as to yield 100% of the radio-activity measured by the ^{51}Cr module as well as ^{125}I (windows: lower $= 0$ upper $= 165$). Thus the radioactivity present as ^{125}I in a mixed sample equals cpm (variable module) – cpm (^{51}Cr module).

3.2.2. Assay

1. Place 100 µL of the doubly-radiolabeled target cells (10^4) in a 1.5-mL microcentrifuge tube labeled "B" and gently mix with either 100 µL of TCM (for spontaneous release) or with various numbers of effector cells ($10^4–10^6$) in 100 µL of TCM (experimental).

2. Centrifuge tubes ($200g$ for 10 min at 22°C) to establish cell-to-cell contact. Incubate for 1–16 h at 37°C. The incubation time will vary depending on the cells being assayed.

3. At the end of the desired incubation period, add 800 µL of ice-cold TCM to each microcentrifuge tube and pellet intact cells and large debris by centrifugation ($200g$ for 10 min at 22°C).

4. Carefully withdraw culture supernatant, place in tube labeled "S," and set aside. The S fraction contains ^{51}Cr and ^{125}I-UdR-labeled DNA fragments released from target cells.

5. Lyse cell pellet in tube B by adding 1 mL of TTE and vortexing vigorously. Microcentrifuge (13,000g; for 10 min at 22°C) to separate intact chromatin from fragmented DNA. Carefully transfer the supernatant to a tube labeled "T." TTE lyses the cells releasing ^{125}I-UdR-labeled DNA fragments and the remaining ^{51}Cr.

6. Quantify the amounts of ^{125}I and ^{51}Cr present in the S, T, and B fractions with a gamma counter.

7. Calculate percentage fragmented DNA (as cpm ^{125}I) for spontaneous and experimental conditions using the following formula:

$$\% \text{ fragmented DNA} = \frac{S + T}{S + T + B} \times 100$$

where S + T is the ^{125}I radioactivity (cpm) present as unsedimented chromatin in the S and T fractions, and S + T + B is the ^{125}I in S, T, and B fractions. Spontaneous DNA fragmentation should not exceed 30% (*see* **Note 7**).

8. Calculate % lysis (as cpm ^{51}Cr release) for spontaneous and experimental conditions using the following formula:

$$\% \text{ lysis} = \frac{S}{S + T} \times 100$$

where S is the radioactivity (cpm) present as ^{51}Cr in the S fraction, and S + T is the radioactivity present as ^{51}Cr in the S and T fractions.

9. Calculate percent specific fragmented DNA or lysis using the following formula:

$$\% \text{ specific fragmented DNA or lysis} = \frac{expt - spont}{100 - spont} \times 100$$

where *expt* is the experimental value (with NK cells) and *spont* is the spontaneous value (with TCM alone) for percent fragmented DNA or lysis as calculated in **steps 7** and **8**.

10. By utilizing cells that are labeled with ^{125}I-UdR or [^{3}H]TdR, but not with ^{51}Cr, it is possible to use this assay to quantify DNA fragmentation alone.

11. Spontaneous DNA fragmentation should not exceed 30% (*see* **Note 7**).

3.3. Quantification of DNA Fragmentation
Using the [^{3}H]TdR Release Assay or JAM Test

In this protocol, as first described by Matzinger *(15)*, target cells labeled in their DNA with [^{3}H]TdR are harvested after the coincubation period with NK cells onto a fiberglass filtermat using a semiautomated cell harvester. DNA that has been fragmented passes through the filtermat whereas intact, chromosome-length, DNA does not. The level of target cell DNA fragmentation is therefore inversely proportional to the amount of [^{3}H]TdR retained on the filtermat.

3.3.1. DNA Radiolabeling With [^{3}H]TdR

1. Subculture target cells to be labeled at 1–4×10^5 cells/mL in TCM (in tissue culture flasks, multiwell plates, or Petri dishes). It is necessary only that the cells be growing well (*see* **Note 7** for further information regarding DNA labeling).
2. Add 0.5–2 µCi/ml [^{3}H]TdR to cells and incubate for 12–18 h at 37°C.
3. Wash cells once with TCM to eliminate unincorporated [^{3}H]TdR (prewarmed to 37°C).
4. Resuspend cells in TCM at 1×10^4 to 1×10^5 cells/mL for assay (*see* **Subheading 3.3.2.**).
5. Determine level of incorporation by measuring radioactivity in 10,000 target cells with a beta scintillation counter. This procedure should yield an incorporation level of 0.2–3 cpm [^{3}H]/cell. If lower levels of [^{3}H]TdR incorporation are observed, the cells may be infected with mycoplasma or may have not been subcultured at a low enough initial density.

3.3.2. Assay

1. Place 100 µL of [^{3}H]TdR-labeled target cells (10^3–10^4) in individual wells of 96-well, U-, or V-bottomed microtiter plates that contain various concentrations of effector cells (10^4–10^6) in 100 µL of TCM. Three wells containing target cells alone are used for the determination of spontaneous DNA fragmentation.
2. Centrifuge the plates ($200g$; 10 min; 22°C) to establish cell-to-cell contact. Incubate for 1–16 h at 37°C. The incubation time will vary depending on the cells being assayed.

3. At the end of the desired incubation period, the plates are harvested using a semiautomated cell harvester and the counts retained on the filtermat are quantified with a beta scintillation counter in the same manner as for proliferation assays employing [^{3}H]TdR.
4. Percent specific DNA fragmentation is calculated using the following formula:

$$\text{\% specific fragmented DNA or lysis} = \frac{spont - expt}{spont} \times 100$$

where *spont* is the [^{3}H] radioactivity (cpm) retained on the filtermat from target cells incubated without NK cells (spontaneous) and *expt* is the [^{3}H] radioactivity (cpm) retained on the filtermat from target cells incubated with NK cells (experimental).

3.4. Qualitative Analysis of Internucleosomal DNA Fragmentation by Agarose Gel Electrophoresis

In this protocol, fragmented target cell DNA is isolated, concentrated, and analyzed by agarose gel electrophoresis. This procedure demonstrates the internucleosomal DNA cleavage associated with apoptosis *(9,14)*, but is not quantitative owing to limitations inherent in DNA recovery and solubilization. Therefore, this method is used to confirm an observation of apoptosis made with the quantitative assays.

Target cells labeled in their DNA with ^{125}I-UdR or [^{3}H]TdR, *but not* labeled cytoplasmically with ^{51}Cr, are incubated with NK cells in microcentrifuge tubes as described in **Subheading 3.2.2.** In general, 1×10^4 ^{125}I-UdR-labeled or 1×10^5 [^{3}H]TdR-labeled targets should be used per condition to insure that sufficient radioactivity is present to detect target cell-derived DNA fragments following electrophoresis.

3.4.1. Assay

1. Place 100 μL of radiolabeled target cells in a 1.5-mL microcentrifuge tube labeled "B" and gently mix with either 100 μL of TCM (for spontaneous release) or with various numbers of effector cells (10^4–10^7) in 100 μL TCM (experimental).

2. Centrifuge tubes (200g; 10 min; 22°C) to establish cell-to-cell contact. Incubate for 1–16 h at 37°C. The incubation time will vary depending on the cells being assayed.

3. At the end of the desired incubation period, add 300 µL of ice-cold PBS to each microcentrifuge tube and pellet intact cells and large debris by centrifugation (200g; 10 min; 22°C).

4. Carefully withdraw culture supernatant, place in tube labeled "S," and set aside. The S fraction contains radiolabeled DNA fragments released from target cells.

5. Lyse cell pellet in B tube by adding 0.5 mL of TTE and vortexing vigorously. Microcentrifuge (13,000g; 10 min; 22°C) to separate intact chromatin from fragmented DNA. Carefully transfer the supernatant to a tube labeled "T." TTE lyses the cells releasing radiolabeled DNA fragments.

6. Add 0.5 m: of TE to the pelleted nuclei and large chromatin in tube B.

7. Add 0.1 mL of ice-cold 5 M NaCl to the approx 0.5-mL solution in tubes S, T, and B. Vortex vigorously and add 0.7 mL of ice-cold isopropanol to each tube. Vortex vigorously and place at –20°C overnight to precipitate DNA.

8. Microcentrifuge 10 min at maximum speed, 4°C. Carefully remove supernatant by rapidly inverting tube or by aspirating with a disposable pipet tip attached to a vacuum line. With tube upside down, use a cotton-tipped swab to remove any drops adhering to walls of tube, taking care not to disturb the DNA pellet. The precipitated DNA and some of the salt form a loose pellet that may be virtually invisible.

9. Half-fill tubes with ice-cold 70% ethanol and microcentrifuge 10 min at maximum speed, 4°C. Carefully remove and discard supernatant as described in **step 3**.

10. Stand tube in an inverted position over absorbent paper to allow as much remaining supernatant as possible to drain away. After 30 min, place tube upright, and allow to air dry 3–4 h.

11. Add 20–50 µL of TE buffer to DNA pellets and incubate for 24–72 h at 37°C. The time required to solubilize the DNA depends on the amount and average size of the DNA present in the sample. Because this protocol is for a qualitative assessment of apoptosis, analysis of the T fractions is of greatest interest. The DNA in the T fractions is small and can be readily solubilized in TE buffer such that it is ready to apply to the agarose gel within 24 h.

12. Add 10X loading buffer to 1X final concentration. Heat 10 min at
    65°C. Immediately apply 10–20 µL of sample to well of a 1% agar-
    ose gel and electrophorese using a standard TBE buffer until bro-
    mphenol blue dye has migrated to approx 2 cm from end of gel.
13. Visualize radiolabeled DNA by autoradiography or by using a
    phosphorimager.
14. This protocol is simple and provides reasonably good results. How-
    ever, the DNA is not thoroughly cleaned of proteins, salts, and deter-
    gents. Thus a more thorough DNA isolation method may be required
    if the results obtained are not satisfactory.

3.5. Flow Cytometric Quantification
of Apoptosis Using TUNEL Staining

Terminal deoxynucleotidal transferase (TdT)-mediated dUTP
nick end-labeling (TUNEL) is a method for detecting apoptotic cells
that exhibit double-stranded DNA fragmentation *(16,17)*. This protocol outlines a method for quantifying TUNEL-positive cells by
flow cytometric analysis. In addition, it is compatible with simultaneous multicolor cell surface staining allowing quantification of
both target and NK cell death by performing TUNEL analysis
immediately following cell surface staining. *See* **Note 6** for further
information regarding the TUNEL assay.

1. Incubate a minimum of 5×10^5 target cells with an appropriate number of effector cells in a 12×75-mm polypropylene tissue culture
   tube in a total volume of 1 mL TCM.
2. At the end of the incubation period, fill the tube with ice-cold PBS
   and centrifuge ($200g$; 10 min; 4°C) to pellet cells. Discard supernatant.
3. Add 1 mL of ice-cold 4% paraformaldehyde (prepared in PBS) and
   tap gently to mix (*see* **Notes 8–11** for further information regarding
   fixation and permeabilization).
4. Incubate 15 min on ice and transfer to a 1.5-mL microcentrifuge tube.
5. Pellet cells by centrifugation ($400g$; 5 min; 4°C), discard supernatant and wash once with 1.5 mL of PBS.
6. Add 200 µL PBS, tap gently to resuspend.
7. Add 1 mL of ice cold 70% ethanol pipetting gently to mix (*see* **Notes
   8** and **11**).

8. Place at −20°C for at least 3 d (*see* **Notes 9–11**).
9. Pellet cells by centrifugation (600*g*; 5 min; 4°C), discard supernatant and wash twice with 1.5 mL of PBS. Be certain that cells are being pelleted efficiently as fixed cells are more difficult to pellet.
10. Remove as much of the supernatant as possible and discard.
11. Add 0.5 mL of TdT reaction buffer, pellet cells, and remove as much supernatant as possible by inverting the tube and touching the lip to a cotton-tipped swab after decanting.
12. Add 50 µL of the TdT/FITC–dUTP reaction mixture. Protect cells from light from this step forward.
13. Incubate for 60 min at 37°C.
14. Add 400 µL of PBS containing 0.1% BSA. Pellet cells by centrifugation (600*g*; 5 min; 4°C).
15. Wash once more with 1.5 mL of PBS, resuspend cells in 0.5 mL of PBS
16. Perform flow cytometric analysis, gating cells on a control sample consisting of target cells incubated alone and treated as described in **steps 1–15** but omitting the TdT from the 50-µL reaction mixture in **step 12**.
17. For multicolor analysis to detect target or effector cells separately, perform immunofluorescence cell surface staining prior to the TUNEL procedure and resuspending cells in PBS. It is important to use PE- or Texas red-, but not allopyocyanin-, conjugated antibodies for cell surface staining as these fluorochromes are not destroyed by the fixation steps used in the TUNEL procedure.
18. A commercial kit can be used in place of this protocol and following the manufacturer's instructions; however, purchasing the individual reagents is much more economical.

3.6. Distinguishing Fas-Mediated From Granule Exocytosis-Mediated Target Cell Apoptosis

NK cells predominantly make use of granule exocytosis to kill their targets; however, it has recently been reported that NK cells can express Fas ligand and kill targets expressing the Fas receptor (**ref. 7,8**; *see* also **Note 3**). It is important to note that some target cells that are sensitive to NK cell-mediated killing, such as YAC-1, are capable of being killed via Fas receptor whereas others, such as K562, are not. Thus it is possible to discriminate between Fas-mediated and granule exocytosis-mediated target cells apoptosis on this basis alone.

With target cells that express the Fas receptor, the contribution of Fas-mediated apoptosis to cytolysis can be determined in three ways. First, the assay can be performed in the presence of EGTA. EGTA chelates extracellular calcium which is required for granule exocytosis- but not Fas-mediated apoptosis *(18)*. Second, Fas-mediated apoptosis can be blocked by soluble Fas-Fc *(18)*. Finally, Fas-mediated apoptosis can be blocked with anti-Fas ligand antibody *(19)*.

When performing assays in the presence of EGTA, it is critical to understand that extracellular calcium is required not only for granule exocytosis but also for receptor-mediated activation of the NK cell to express Fas ligand *(20)*. To overcome this potential underestimation of NK cell activity, effector cell populations can be treated with PMA plus Ionomycin prior to addition to the assay. Pretreatment with PMA + Ionomycin results in expression of Fas ligand by lymphocytes that can then kill target cells expressing Fas receptor in the presence of EGTA.

The strategies for discriminating between Fas- and granule exocytosis-mediated apoptosis can be applied to any of the protocols described above in **Subheadings 3.1.–3.5.**

3.6.1. Using PMA and Ionomycin to Stimulate NK Cell Effector Populations to Express Fas Ligand

This subprotocol describes a method for treating NK cell effector populations with PMA and Ionomycin to induce expression of Fas ligand prior to coincubation with target cells.

1. Pellet NK cell effectors and resuspend in 2 mL of TCM in 12×75 mm polypropylene tissue culture tube.
2. Add 2 µL of 1000× PMA stock solution (final conc. = 5 ng/mL). Vortex thoroughly to mix. PMA is only slightly soluble in aqueous solutions so immediate vortexing is required.
3. Add 2 µL of 100× Ionomycin stock solution (final conc. = 500 ng/mL). Vortex thoroughly to mix. Ionomycin is only slightly soluble in aqueous solutions so immediate vortexing is required.
4. Incubate for 2–3 h at 37°C.

5. Wash cells twice with TCM and resuspend in TCM at the appropriate concentration for use in cytotoxicity and/or apoptosis assays as described in **Subheadings 3.1.–3.5.**

3.6.2. Using EGTA to Inhibit Granule Exocytosis-Mediated Apoptosis

1. Resuspend target cells at the appropriate concentration for use in cytotoxicity and/or apoptosis assays as described in **Subheadings 3.1.–3.5.** in TCM alone or in TCM containing EGTA and $MgCl_2$ such that the final concentration of EGTA and $MgCl_2$ present in the assay is 5 mM and 10 mM, respectively.
2. TCM containing EGTA and $MgCl_2$ is prepared fresh using the EGTA/$MgCl_2$ stock solution described in **Subheading 2.**
3. Comparing results obtained with assays run in the presence (granule exocytosis-mediated and Fas-mediated apoptosis) vs absence (Fas-mediated apoptosis only) of extracellular calcium gives an estimation of the relative contribution of each pathway to the cytolytic potential of a given NK cell effector population.

3.6.3. Using Soluble Fas–F_c and Anti-Fas Ligand Antibody to Inhibit Fas-Mediated Apoptosis

Soluble Fas-F_c and anti-Fas ligand antibody are added to assays at a final concentration of 10 µg/mL and 5 µg/mL to block Fas-mediated apoptosis.

4. Notes

1. **Figure 1** shows a schematic representation of how NK cell-induced target cell apoptosis is thought to proceed. Despite an enormous amount of progress in the last few years, the mechanism of apoptosis remains poorly understood at this time, even in systems that have been extensively studied. As discussed in greater detail below, it is impossible to predict whether DNA fragmentation in a given cell type undergoing apoptosis will involve double-stranded cleavage, single-stranded nicking, or any fragmentation. Similarly, not all cells contain the same types of caspases and not all caspases are activated by different inducers in the same cell type. Therefore,

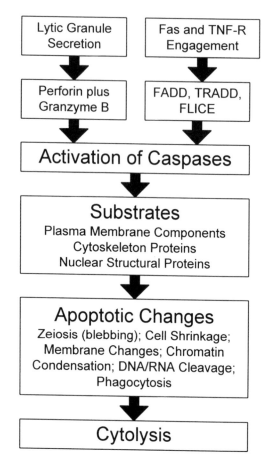

Fig. 1. Schematic representation of NK cell-induced apoptosis. In brief, NK cells can use two mechanisms to induce apoptosis in their targets. Mechanism 1 involves secretion of lytic granules; apoptosis is induced by granzymes in combination with perforin. Granule-mediated exocytosis requires engagement of the NK cell receptor and extracellular calcium is required for granule secretion as well as perforin polymerization. Mechanism 2 involves expression of Fas ligand by the NK cell which can induce apoptosis in target cells expressing functional Fas receptor. Expression of Fas ligand requires engagement of the NK cell receptor and extracellular calcium; however, once Fas ligand is expressed, induction of apoptosis proceeds in the absence of extracellular calcium. Both mechanisms lead to activation of caspases and cleavage of substrates resulting in biochemical changes in the target cell encompassing apoptic cell death.

when examining NK cell-mediated killing involving a cell type or effector cell population or molecule that has not been reported, an empirical approach must be undertaken.

2. It is important to understand that apoptosis was first—and probably is still best—described morphologically. This point cannot be stressed enough to investigators who are new to the field and who may consider morphological studies too subjective and/or cumbersome. Our own experience indicates that apoptosis assays based on biochemical changes or flow cytometric analyses should always be backed up with morphological studies. It is extremely important to acquaint oneself with normal chromatin distribution in the cell line being investigated; different cell types may present strikingly different patterns of "normal" chromatin organization. Similarly, it is a good idea to obtain experience in recognizing apoptotic nuclei in the cells under investigation. A convenient way to induce apoptosis in most cell types is to treat them with 20–100 μM valinomycin (Sigma) for 1–24 h *(10,21)*. Valinomycin is a potassium ionophore that induces apoptosis rapidly in most cultured cell lines and is useful as a positive control. It is prepared as a 5 mM stock solution in EtOH which can be stored indefinitely at –20°C or –70°C. It is important to vortex samples thoroughly after addition of valinomycin and to not add the compound in such a way that the final concentration of EtOH in the cell cultures exceeds 2% (v/v). **Figure 2** shows the typical change in nuclear chromatin morphology that accompanies apoptosis induced by valinomycin. As a positive control for necrosis, cells can be heated to 43–45°C for 2 h.

3. NK cells can use two separate mechanisms to induce apoptosis. The first, and predominant, mechanism involves secretion of lytic granules that contain perforin and a variety of enzymes, the most important of which appears to be granzyme B (*see* also Chapter 13). Granzyme B is a potent activator of many of the caspases that have been identified in apoptosis. In purified form, the ability of granzyme B to induce apoptosis requires sublethal concentrations of perforin *(6)*. Thus it is important to understand that some NK cell populations may induce necrosis before granzyme B is able to induce apoptosis, particularly those effector cell populations that express high amounts of perforin. This could explain the conflicting reports concerning the ability of NK cells, or lytic granules derived from these cells, to induce apoptosis vs necrosis *(1,2,22)*. The second mechanism involves expression of Fas ligand, and perhaps tumor necrosis factor (TNF),

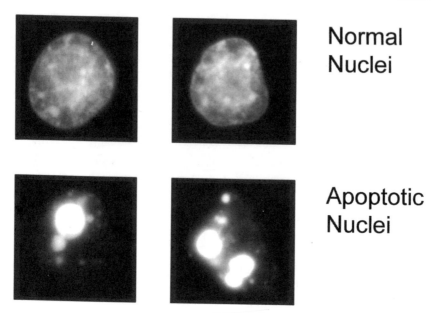

Normal Nuclei

Apoptotic Nuclei

Fig. 2. Nuclear changes typical of apoptotic cell death. Murine leukemia cells were incubated alone (*top panels*) or with 50 μm valinomycin (*bottom panels*) for 2 h at 37°C. Cells were stained with Hoechst 33342 and observed by UV microscopy. Normal nuclei have "structure;" variations in fluorescent intensity reflect the distribution of euchromatin and heterochromatin. Apoptotic nuclei, in contrast, have highly condensed chromatin that is uniformly fluorescent. This can take the form of crescents around the periphery of the nucleus, or the entire chromatin can be present as one or a group of featureless, bright spherical beads.

by activated NK cells. This pathway is probably of little consequence in the killing of targets directly recognized by NK cells as it requires much longer to initiate than the perforin/granzyme B pathway. However, it could play a role in killing bystander cells that express functional Fas receptor and should be considered. Potential targets for bystander killing by NK cells in vivo could be activated T and B cells, macrophages, and certain subsets of bone marrow progenitors (*23,26*).

4. Although the demonstration of extensive DNA fragmentation and a nucleosome ladder have been used as absolute evidence for apoptosis,

this conclusion is not warranted as there are now numerous examples of cells that do not degrade their DNA to oligonucleosomes during apoptosis. For example, some cells undergo random double-stranded DNA cleavage only every 50–300 kilobases; this event may actually precede internucleosomal cleavage in most cell types undergoing apoptosis. Nevertheless, the large fragments cannot be detected in conventional sedimentation or agarose gel electrophoresis assays but instead are observed using pulsed-field electrophoresis or by neutral sucrose-gradient density centrifugation *(27)*. Once again, it is important to note that atypical chromatin cleavage was observed using regimens that induced typical apoptosis in related cells; the changes were not observed if the cells were killed via necrosis. Cells that are necrotic may also undergo a minor amount of DNA fragmentation although this is typically not internucleosomal in nature.

5. CAUTION: Acridine orange, ethidium bromide, and propidium iodide have been found by the Ames test to be highly mutagenic and should be handled with care.

6. One practical drawback of the TUNEL technique is that it is much more expensive to perform than other methods (e.g., staining with AO + EB costs approx $0.10 per sample vs as much as $8.00 per sample using TUNEL). It is highly recommended that TdT as well as FITC-dUTP be titrated by the investigator to avoid using them in excessively high concentrations. The wash step with TdT buffer (**Subheading 3.5., step 11**) optimizes conditions for TdT and is critical for achieving good end labeling with dUTP and TdT at low concentrations.

7. Radiolabeling DNA may lead to high (>30%) values for spontaneous DNA fragmentation which may not be associated with apoptotic morphology. If the cells being tested are infected with mycoplasma, the mycoplasma may incorporate the radiolabel, which will appear in the S or T fractions during quantification, leading to an apparently high spontaneous DNA fragmentation even though the cells appear healthy. In general, high levels of spontaneous DNA fragmentation suggest that the cells are being adversely affected. This finding should not be ignored and adequate measures should be taken to obtain lower values of spontaneous fragmentation and apoptosis.

8. The final ethanol concentration used to permeabilize the cells should be 60–70% as higher concentrations will cause excessive shrinkage and difficulties in obtaining a cell pellet as well as acquiring data in the flow cytometer.

9. For optimal results it is recommended that the cells be stored in 70% ethanol for at least 3 d (**Subheading 3.5., step 8**) before continuing with end-labeling.
10. In some cells reversing **steps 3–8** such that cells are fixed/permeabilized with 70% ethanol prior to fixation with paraformaldehyde may provide better results.
11. To avoid excessive clumping of cells, it is recommended that paraformaldehyde and ethanol be added slowly as the cells are gently mixed.

References

1. Duke, R. C., Cohen, J. J., and Chervenak, R. (1986) Differences in target cell DNA fragmentation induced by mouse cytotoxic T cells and natural killer cells. *J. Immunol.* **137,** 1442–1447.
2. Zychlinsky, A., Zheng, L. M., Liu, C. C., and Young, J. D. E. (1991) Cytotoxic lymphocytes induce both apoptosis and necrosis in target cells. *J. Immunol.* **146,** 393–400.
3. Shi, L., Kam, C. M., Powers, J. C., Aebersold, R., and Greenberg, A. H. (1992) Purification of three cytotoxic lymphocyte granule serine proteases that induce apoptosis through distinct substrate and target cell interactions. *J. Exp. Med.* **176,** 1521–1529.
4. Kagi, D., Ledermann, B., Burki, K., Seiler, P., Odermatt, B., Olsen, K. J., Podack, E. R., Zinkernagel, R. M., and Hengartner, H. (1994) Cytotoxicity mediated by T cells and natural killer cells is greatly impaired in perforin-deficient mice. *Nature* **369,** 31–37.
5. Duke, R. C., Ojcius, D., and Young, J. D. -E. (1996) Cell suicide in health and disease. *Sci. Am.* **275,** 80–87.
6. Greenberg, A. H. (1996) Granzyme B-induced apoptosis. *Adv. Exp. Med. Biol.* **406,** 219–228.
7. Arase, H., Arase, N., and Saito, T. (1995) Fas-mediated cytotoxicity by freshly isolated natural killer cells. *J. Exp. Med.* **181,** 1235–1238.
8. Eischen, C. M. and Leibson, P. J. (1997) Role for NK-cell-associated Fas ligand in cell-mediated cytotoxicity and apoptosis. *Res. Immunol.* **148,** 164–169.
9. Duke, R. C., Chervenak, R., and Cohen, J. J. (1983) Endogenous endonuclease-induced DNA fragmentation, an early event in cell-mediated cytolysis. *Proc. Natl. Acad. Sci. USA.* **80,** 6351–6355.
10. Duke, R. C., Zulauf, R., Nash, P. B., Young, J. D. -E., and Ojcius, D. (1994) Cytolysis mediated by ionophores and pore-forming agents, role of intracellular calcium in apoptosis. *FASEB J.* **8,** 237–246.

11. Duke, R. C., Persechini, P. M., Chang, S., Liu, C. -C., Cohen, J. J., and Young, J. D. -E. (1989) Purified perforin induces target cell lysis but not DNA fragmentation. *J. Exp. Med.* **170,** 1451–1456.

12. Kagi, D., Ledermann, B., Burki, K., Zinkernagel, R. M., and Hengartner, H. (1996) Molecular mechanisms of lymphocyte-mediated cytotoxicity and their role in immunological protection and pathogenesis in vivo. *Annu. Rev. Immunol.* **14,** 207–232.

13. Liu, C. C., Persechini, P. M., and Young, J. D. (1995) Perforin and lymphocyte-mediated cytolysis. *Imm. Rev.* **146,** 145–175.

14. Wyllie, A. H. (1980) Glucocorticoid-induced thymocyte apoptosis is associated with endogenous endonuclease activation. *Nature* **284,** 555–557.

15. Matzinger, P. (1991) The JAM test, a simple assay for DNA fragmentation and cell death. *J. Immunol. Methods* **145,** 185–192.

16. Gavrieli, Y., Sherman, Y., and Ben-Sasson, S. A. (1992) Identification of programmed cell death in situ via specific labeling of nuclear DNA fragmentation. *J. Cell Biol.* **119,** 493–501.

17. Gorczyca, W., Gong, J., and Darzynkiewicz, Z. (1993) Detection of DNA strand breaks in individual apoptotic cells by the in situ terminal deoxynucleotidyl transferase and nick translation assays. *Cancer Res.* **53,** 1945–1951.

18. Rouvier, E., Luiciani, M. F., and Golstein, P. 1993. Fas involvement in Ca(2+)-independent T cell-mediated cytotoxicity. *J. Exp. Med.* **177,** 195–200.

19. Tanaka, M., Suda, T., Haze, K., Nakamura, N., Sato, K., Kimura, F., Motoyoshi, K., Mizuki, M., Tagawa, S., Ohga, S., Hatake, K., Drummond, A. H., and Nagata, S. (1996) Fas ligand in human serum. *Nat. Med.* **2,** 317–322.

20. Vignaux, F., Vivier, E., Malissen, B., Depraetere, V., Nagata S., and Golstein, P. (1995) TCR/CD3 coupling to Fas-based cytotoxicity. *J. Exp. Med.* **181,** 781–786.

21. Ojcius, D. M., Zychlinski, A., Zheng, L. M. and Young, J. D. Y. (1991) Ionophore-induced apoptosis, role of DNA fragmentation and calcium fluxes. *Exp. Cell Res.* **197,** 43–49.

22. Duke, R. C., Sellins, K. S., and Cohen, J. J. (1988) Cytotoxic lymphocyte-derived lytic granules do not induce DNA fragmentation in target cells. *J. Immunol.* **141,** 2141–2145.

23. Miyawaki, T., Uehara, T., Nibu, R., Tsuji, T., Yachie, A., Yonehara, S., and Taniguchi, N. (1992) Differential expression of apoptosis-

related Fas antigen on lymphocyte subpopulations in human peripheral blood. *J. Immunol.* **149,** 3753–3758.

24. Leithauser, F., Dhein, J., Mechtersheimer, G., Koretz, K., Bruderlein, S., Henne, C., Schmidt, A., Debatin, K. M., Krammer, P. H., and Moller, P. (1993) Constitutive and induced expression of APO-1, a new member of the nerve growth factor/tumor necrosis factor receptor superfamily, in normal and neoplastic cells. *Lab. Invest.* **69,** 415–427.

25. Maciejewski, J., Selleri, C., Anderson, S., and Young, N. S. (1995) Fas antigen expression on CD34+ human marrow cells is induced by interferon gamma and tumor necrosis factor alpha and potentiates cytokine-mediated hematopoietic suppression in vitro. *Blood* **85,** 3183–3190.

26. George, T., Yu, Y. Y., Liu, J., Davenport, C., Lemieux, S., Stoneman, E., Mathew, P. A., Kumar, V., and Bennett, M. (1997) Allorecognition by murine natural killer cells, lysis of T-lymphoblasts and rejection of bone-marrow grafts. *Imm. Rev.* **155,** 29–40.

27. Sellins, K. S. and Cohen, J. J. (1991) Cytotoxic T lymphocytes induce different types of DNA damage in target cells of different origins. *J. Immunol.* **147,** 795–803.

5

Detecting and Quantifying Apoptosis in Tissue Sections

Vicki Save, Peter A. Hall, and Philip J. Coates

Summary

Various techniques exist for the identification of apoptosis in tissue sections or intact cells. The use of simple morphology, electron microscopy, DNA-end labeling techniques, and immunochemical methods are reviewed, with a particular emphasis on *in situ* end-labeling. The analysis of apoptotic cells and methods for their quantification are also discussed.

Key Words: Apoptosis; ISEL; TUNEL; immunohistochemistry; histology; cytology.

1. Introduction

The realization that apoptosis represents a critical element in cell number control in physiological and pathological situations has been well reviewed *(1–4)*. In addition, many of the effects of chemo- and radiotherapeutic agents are mediated by apoptosis *(2,3)*, making the identification and quantitation of apoptosis in normal and pathological tissues of great importance. The seminal work of Kerr, Wyllie, and Currie *(5)*, building on the earlier observations of Glucksman *(6)* and Saunders *(7)*, should be read by those interested in assaying apoptosis in tissues because of the excellent photomicrographs that document the morphological features of the process.

From: *Methods in Molecular Biology, vol. 282: Apoptosis Methods and Protocols*
Edited by: H. J. M. Brady © Humana Press Inc., Totowa, NJ

This is important, because despite considerable progress in the understanding of the mechanistic basis of apoptosis, morphological analysis remains unquestionably the "gold standard" for its assessment and quantitation.

Apoptosis is a regulated and active process. Although a diverse range of insults and physiological events can lead to apoptosis, the process is remarkably stereotyped, with a program of activities leading to the final morphological events that are similar throughout phylogeny and may be recapitulated in most (if not all) cell types. Mounting data indicate that much of the machinery for the implementation of the apoptotic response is "hard-wired" in cells, being present all the time but kept in an "off" state but rapidly recruited into an "on" state if needed. Consequently, and despite much effort, there remain few biochemical markers of apoptosis that are specific for this complex regulated process. Similarly, although many potential regulators of apoptosis are described, critical examination of the available data indicates that there is little consensus on their value as markers of apoptosis.

A critical point for the quantitation of apoptosis is that, irrespective of the initiating insult, the time course of apoptosis is very fast *(8,9)*. Moreover, the clearance of the resultant debris (either by professional phagocytes or bystander "amateur" phagocytes) is rapid. Coles et al. *(10)* suggested that clearance times of less than 1 h were typical. The rapid nature of apoptosis means that in any static analysis a very small number of apoptotic cells observed at a given instant might, in fact, reflect a very considerable contribution to cell turnover. Dramatic evidence for this came from studies of the physiological contribution of apoptosis in renal development *(10)* and in the steady-state regulation of intestinal epithelial populations *(11)*. Although the process of apoptosis and its clearance is (as far as we can tell) always rapid, there may be some variation between cell types or in relation to different insults. This has important implications for the quantitation of apoptosis, as highlighted by Potten *(12)*.

Given its contribution to cell turnover in physiological, pathological, and toxicological situations, it is important to be able to identify and quantitate the process of apoptosis in cells and in tissues. Ideally, one would like a technique that is sensitive and highly selective for apoptosis, as well as being easily applicable to routinely prepared tissue sections. Here, we review the various options that can be used to identify and quantitate apoptosis, concentrating on methods applicable to tissues sections. A detailed method for the identification of apoptosis using *in situ* end-labeling of DNA fragments (terminal deoxynucleotidyl transferase-mediated dUTP nick end-labeling [TUNEL]/*in situ* end-labeling [ISEL]) is given.

1.1. Demonstration of Apoptosis: Comparison of Methods

1.1.1. Morphology

Apoptosis is defined by a series of morphological changes *(5,13,14)*. The classical features are best seen by electron microscopy but can be observed at the light microscopic level using nucleic acid-binding dyes, such as hematoxylin, acridine orange, or propidium iodide *(5,10,11)*. The first signs of apoptotic cell death are a condensation of the nuclear material, with a marked accumulation of densely stained chromatin, typically at the edge of the nucleus. This is accompanied by cell shrinkage. Cytoplasmic blebs appear on the cell surface, and the cell detaches from its neighbors. The nuclear outline often becomes highly folded and the nucleus breaks up, with discrete fragments dispersing throughout the cytoplasm. Eventually, the cells themselves fragment, with the formation of a number of membrane-bound apoptotic bodies. Cells undergoing apoptosis are rapidly phagocytosed by surrounding cells, which are not necessarily derived from the mononuclear phagocytic system. Therefore, the most common sign of apoptosis in a tissue section is the presence of apoptotic bodies inside other cells. Apoptotic bodies have a diverse appearance, particularly in regard to their size. They are generally oval or round in shape and are most easily recognized when they contain large amounts of homogeneous, condensed chromatin. The morphological features of apoptosis have been extensively reviewed, and plentiful illustrations of both the light and electron microscopic appearances are provided in these publications *(5,10,11,13,14)*.

1.1.2. DNA Fragmentation Assays

A characteristic feature of apoptosis is DNA fragmentation. Wyllie *(15)* described nucleosomal fragmentation, as seen in agarose electrophoresis, as a distinct "ladder" of DNA bands representing multiples of 180–200 bp. This DNA ladder correlates with the early morphological signs of apoptosis *(16)*, and has been widely used as a distinctive marker of the process. DNA laddering is not seen in cells that have undergone necrosis, which show a random fragmentation pattern leading to smears on agarose electrophoresis *(15)*. However, the method cannot be applied (in this form) to tissue sections. In some cases, DNA fragmentation appears to be delayed *(17)*, and cells undergoing apoptosis may show more limited DNA degradation with the formation of 300- or 50-kb fragments *(18)*. These fragments are thought to represent the release of loops of chromatin from their attachment points on the nuclear scaffold and require pulsed field gel electrophoresis to separate the large DNA fragments.

1.1.3. ISEL or TUNEL

The property of DNA fragmentation in apoptosis can be used to identify cells undergoing this process using enzymes that add labeled nucleotides to the DNA ends *(19–23)*. The labeled nucleotides can then be identified by methods akin to immunohistochemistry. These techniques were originally termed TUNEL but are also referred to as ISEL. Strictly, the different names relate to the different enzymes used, and although there are theoretical and some practical differences, the similarity of technique and result make the names essentially interchangeable. A comparison of the methodologies using terminal deoxynucleotidyl transferase (TdT) (TUNEL) and Klenow fragment of DNA polymerase (ISEL) was reported by Mundle et al. *(24)*, who demonstrated that TUNEL appeared more sensitive than ISEL. This is because TdT can label 3' recessed, 5' recessed, or blunt ends of DNA, whereas ISEL labels only those with 3' recessed ends. All three types of DNA end are seen in apoptosis, and thus, in principle, TdT-based methods should be more sensitive than Klenow polymerase methods. Despite this, in practice ISEL and TUNEL appear functionally interchangeable.

Irrespective of the enzyme used, a variety of labels can be used, including radioactive nucleotide triphosphates. However, methods based on the use of nonisotopic labels are superior for a variety of reasons, including ease of use, stability, simplicity and speed of detection, and the increased resolution obtained. Using this approach, it has been clearly shown that the amount and distribution of labeled cells is closely correlated with the amount and distribution of cells known to be undergoing apoptosis using other methods (20–23). The method can be modified for fluorescence detection *in situ* or by flow cytometry or for detection at the light or ultrastructural levels (23,25). In addition, the combination of immunocytochemistry or *in situ* hybridization with TUNEL/ISEL allows the identification of the particular cell types undergoing apoptosis or can be used to measure phenotypic changes in apoptotic cells (11,23,26,27).

Protocols for TUNEL/ISEL are given later based on our experience (11,22,28,29; *see* **Note 1**). A number of variables must be considered when performing the technique. The staining results depend on the rapidity and extent of fixation, the extent of proteolytic digestion, the extent of incorporation of labeled nucleotide triphosphate, and the sensitivity of detection of the label.

1.1.4. In Situ DNA Ligation

This technique can be considered as a variation of TUNEL/ISEL in that it uses labeled nucleotides to identify DNA breaks in apoptotic cells. However, the technique varies significantly from the other methods in that biotin-labeled hairpin oligonucleotides are covalently joined onto double-strand breaks in DNA using DNA ligase (30,31). The method can detect either blunt ends or protruding ends by the design of the oligonucleotide and has recently been used to demonstrate differences in the type of breaks seen in apoptotic vs necrotic cells (32). A detailed discussion of the method and protocols are given elsewhere in this series (31).

1.1.5. Immunohistochemical Methods

In recent years, a large number of antibodies have been marketed for apoptosis research. Many of these recognize proteins that are involved in signaling the apoptotic response and are neither universal nor specific. Indeed, because apoptosis is a consequence of the activation of pre-existing mechanisms within a cell, there has been relatively little progress in the identification and use of antigens whose expression is specific for this form of cell death. However, some progress has been made recently based on increased knowledge of the common processes involved in apoptosis. The process of apoptosis involves caspases, which are proteolytic enzymes that cleave a wide range of intracellular proteins. Because caspases are present in normal cells as inactive precursors, the identification of these proteins is not indicative of apoptosis. However, antibodies that specifically recognize active caspases are theoretically good markers of the apoptotic process. For example, the identification and quantification of cells containing activated caspase 3, an executioner caspase, show a good correlation with quantification of morphological apoptosis and TUNEL *(33,34)*. A similar approach involves the specific identification of cleaved protein substrates of caspases. For example, cytokeratin 18 cleavage during apoptosis reveals an epitope recognized by the M30 monoclonal antibody that correlates with other apoptotic measurements *(34,35)*. Although both of these methods appear useful and are becoming more widely used, they will not be universally applicable; the dependence of apoptosis on caspase 3 is both cell-type and stimulus-specific *(36,37)*, and cytokeratin 18 cleavage can only identify apoptosis in cells that contain cytokeratin 18 (i.e., its use is restricted to epithelial subpopulations). Furthermore, caspase activation is associated with normal cellular differentiation functions *(38)* and so will not be specific for apoptosis in all tissues. In addition, it must be realized that these methods identify the early stages of apoptosis, before the characteristic nuclear changes occur, and will therefore be expected to identify different populations of cells undergoing apoptosis. Another immunological approach is to use antibodies that bind specifically to single-stranded DNA, and such reagents appear to be useful markers of apoptotic cells in many instances, discriminating apoptosis from necrosis and correlating with TUNEL/ISEL measurements *(39,40)*. However, single-stranded DNA arises naturally during other cellular processes and after exposure to endogenous or exogenous genotoxic agents. Thus, as with TUNEL/ISEL, this approach is unlikely to be specific for apoptosis under all physiological and pathological conditions. Expression of clusterin (also known as TRPM-2, or SOP-2) has been correlated with apoptosis, but this protein is neither a universal nor specific marker *(41)*. Tissue transglutaminase is expressed during apoptosis and has been widely used as an immunological marker, although again the enzyme is not always induced *(42)*. Finally, apoptosis can be demonstrated by *in situ* hybridization *(43)*, although the utility of this method remains to be established.

1.1.6. Conclusion

Although we have a burgeoning knowledge of the mechanistic basis of apoptosis and an increasing recognition of its contribution to both physiological and pathological processes, the ability to quantitate objectively this process remains poorly developed. It is the view of these authors that if apoptosis is to be assessed in histological material, then there is no escape from meticulous and painstaking microscopy coupled with rigorous and meticulous quantitation. Whether this is the quantitation of morphologically defined events (the "gold standard"), of TUNEL/ISEL-defined events, or of immunochemically identified cells is in our view a relatively unimportant issue; they will correlate very well in most circumstances, but all have inherent problems. Arguably, the morphological approach is perhaps more satisfactory, but, curiously, the additional complexity of other techniques attracts many workers! In the end, the choice of method will depend on the experience of the researcher in histological analysis and microscopy.

2. Materials
2.1. Materials in Common to ISEL and TUNEL

1. Tissues and cells: *In situ* end-labeling techniques can be applied to cells grown in culture, to frozen tissue sections, or to sections of

formalin-fixed, paraffin-embedded material found in surgical pathology archives. Remember that apoptosis of adherent tissue culture cells will often cause the cells to detach—these cells can be collected and analyzed on cytospin preparations.

2. Glass slides: Slides coated with poly-L-lysine or other adhesive are available from most histology suppliers.

3. Methanol/acetone fixative: 50% methanol, 50% acetone. Store at −20°C in a spark-free freezer. Stable for many months.

4. Paraformaldehyde fixative: In a fume hood, dissolve 4 g of paraformaldehyde in 80 mL of water with gentle heating and the addition of 1 M NaOH until the powder dissolves. Make up to 90 mL with distilled water and add 10 mL of 10× phosphate-buffered saline (PBS). Store at 4°C and use within 24 h.

5. Xylene or nontoxic xylene substitute for dewaxing paraffin-embedded sections (available from all histology suppliers).

6. 10× PBS: 1 L consists of 80 g NaCl, 2 g KCl, 14 g Na_2HPO_4, and 2 g KH_2PO_4. Adjust pH to 7.4 with HCL or NaOH as appropriate. Filter-sterilize or autoclave. Store at room temperature.

7. Proteinase K: A stock solution of proteinase K (Sigma, Poole, Dorset, UK; P2308) is prepared by dissolving 10 mg of enzyme in 1 mL of sterile distilled water to give a 10 mg/mL concentration. The enzyme is aliquoted and stored frozen. The aliquots should not be thawed and re-frozen more than twice. For use, the stock solution is diluted into sterile 50 mM Tris-HCl, pH 8.0, and 1 mM ethylenediaminetetraacetic acid.

8. Ethanol: 100%; 95%, 90%, and 70% with distilled water.

9. Methanol.

10. Hydrogen peroxide (100 volumes; 30%).

11. Horseradish peroxidase-conjugated avidin (e.g., Dako Ltd., P364) is made freshly by diluting 1/200 into PBS containing 1% bovine serum albumin (Sigma A9647).

12. Diaminobenzidine (DAB): *see* **Note 2**. In a fume hood, weigh out diaminobenzidine tetrahydrochloride (Sigma D5637) and dissolve to a concentration of 10 mg/mL in PBS. Aliquot in 1-mL samples and store frozen at −20°C. For use, add 1 mL of DAB to 20 mL of PBS and 20 mL of 10 mM imidazole, pH 7.4, and add 30 µL H_2O_2. Use immediately.

13. Mayer's hematoxylin.

2.2. ISEL

1. *In situ* end-labeling buffer: 0.01 mM each of biotin, dATP, dCTP, dGTP, and dTTP in 50 mM Tris-HCl, pH 7.5; 5 mM MgCl$_2$, 10 mM 2-mercaptoethanol, 0.005% bovine serum albumin (molecular biology grade, e.g., Sigma B2518), and 5 units/mL Klenow fragment of DNA polymerase I. The labeling solution without polymerase can be prepared in bulk and stored frozen in aliquots, but the DNA polymerase must be added immediately before use. The biotinylated nucleotide can be obtained from Invitrogen Life Technologies Ltd. (cat. no. 19534-016). Deoxynucleotide triphosphates can be purchased as a set from Roche (cat. no. 1277 049).
2. Klenow fragment of DNA polymerase I, 2000 units/mL (e.g., Roche cat. no. 1008 404).

2.3. TUNEL

1. TUNEL buffer: 30 mM Tris-HCl, pH 7.2, 140 mM sodium cacodylate, 1 mM CoCl$_2$ (supplied with the enzyme). Add 0.1 mM biotin-dUTP (Roche cat no. 1093070).
2. TdT enzyme, 25,000 units/mL (Roche cat. no. 220582).

3. Methods

3.1. Fixation

Frozen sections, cytospins, or tissue-culture cells grown on slides or coverslips may be fixed by immersion for 10 min in methanol/acetone at –20°C, followed by air drying. Alternatively, freshly prepared 4% paraformaldehyde can be used. Fixation with the latter is for 10 min at 4°C, and slides are then rinsed in PBS. Tissues or cells that are to be used for TUNEL/ISEL should be fixed as quickly as possible because delay causes significant artifacts.

Importantly, it has been shown that TUNEL/ISEL can be performed on sections of archival tissues after formalin fixation and storage as wax blocks. When using tissue sections or cells that have been fixed in formaldehyde, it is necessary to use a protease to break some of the crosslinks formed between proteins and thereby allow access of the reagents to the degraded DNA. The extent of digestion varies with the extent of fixation and harsh digestion leads to some nonspecific staining, whereas too little digestion results in a decrease in the intensity of staining of apoptotic cells. Proteolytic digestion is not required after fixation in acetone/methanol.

3.2. ISEL (see **Notes 1** *and* **3**)

1. Paraffin sections (3–4 μm) are mounted on adhesive-coated glass slides and allowed to dry overnight to ensure adherence (*see* **Notes 4 and 5** for other samples).
2. Paraffin-embedded sections are dewaxed by immersion in xylene or xylene substitute for 10 min, followed by a further 10 min in fresh dewaxing agent.
3. Rinse sections in two changes of 100% ethanol and air dry.
4. For formaldehyde-fixed sections/cells, digest with proteinase K diluted to a final concentration of 10 μg/mL for 30 min at 37°C (*see* **Note 6**).
5. Wash in sterile distilled water three times, rinse in 70, 90, and 95% alcohol and air-dry.
6. Prepare 40–60 μL ISEL (or TUNEL) buffer for each section and keep on ice (*see* **Notes 7–11**).
7. Carefully pipet the mixture over the tissue section and place a cleaned glass coverslip over the section to prevent evaporation.
8. Incubate at 37°C for 1 h in a moist chamber.
9. Terminate the reaction by washing sections in distilled water three times, being careful not to scratch the tissue surface when removing the coverslip.
10. Block endogenous peroxidase activity with methanol containing 0.5% H_2O_2 (100 volumes) for 30 min (*see* **Note 12**). Wash three times in distilled water and once in PBS, pH 7.4 (5 min each).
11. Incubate with freshly diluted horseradish peroxidase-conjugated avidin for 30 min at room temperature (*see* **Note 8**).
12. Wash in PBS three times (5 min each) and develop in diaminobenzidine-H_2O_2 for 10 min at room temperature (*see* **Note 2**). Wash in water; lightly counterstain with hematoxylin; dehydrate by immersion in 70%, 90%, 95%, and two changes of 100% ethanol (5 min each); clear in xylene or xylene substitute; and mount in resin.
13. View under a light microscope, where apoptotic nuclei are stained brown and non-apoptotic cell nuclei appear blue (*see* **Note 13**).

3.3. Quantitation

Quantitation is a real problem! Much of the currently available literature on apoptosis is problematic because of inadequate quantitation procedures. The application of stereological and morphometric principles to quantitation in histology is difficult: many authors get around this problem by ignoring it. Moreover, many texts on the subject are at best impenetrable. The use of flow cytometric methods has the advantages of objective assessment of large numbers of events (>10^4 typically): those wishing to use *in situ* techniques will not be easily able to match that but will obtain valuable information relating to microanatomical variation, which may be of fundamental biological importance. It is worth restating that not only is the ability of cells to proliferate and differentiate regulated by position (anatomical and within cellular hierarchies) but so too is apoptosis (*11*). The critical issues are simply stated:

1. The confidence that can be placed on the data depends upon the effort and rigor invested in its generation.
2. Where the levels of apoptosis are low (the usual state of affairs), very large numbers of events must be quantitated for accuracy.
3. Methods based on semiquantitative approaches, the use of high-power fields as denominator, the failure to define reproducibility of assessment, and methods that do not consider the heterogeneity implicit in biological samples are to be deprecated.

Given this, what can be done?

3.3.1. What to Count?

ISEL- or TUNEL-stained samples will make the microscopic assessment easier and perhaps more objective, particularly if the observer is not an experienced microscopist. However, these methods may underestimate the true number of apoptotic bodies and may also be influenced by artefacts arising from variable fixation and nonapoptotic processes (*11,22,28,44*). The former may not matter because the "error" will be systematic and the same in all samples. The latter is significant but can be minimized by careful control of sample handling. In the hands of an experienced microscopist then, fluorescent dye-based assays *(10)*, or even simple hematoxylin, will be useful *(11)*. Double labeling may be useful in some circumstances, and ISEL methods can be combined with both immunocytochemical techniques (*see*, e.g., **refs. *11*** and ***26***) and with *in situ* hybridization methods (*see*, e.g., **ref. *27***) to good effect.

3.3.2. How Many Events to Count?

Pragmatism must be the key word and there must be a compromise among the quality of the data, the time taken to generate that data, and the importance of the question. Although statistical approaches can be used to determine how many events must be assessed (*see* **ref. *45*** for a discussion), the authors favor an experimental approach based on the generation of a "wandering mean" in a small number of representative examples from the population of cases to be studied. To generate these data, the following procedure should be undertaken. Count the number of events (TUNEL/ISEL-positive or apoptotic bodies) and the total number of relevant cells in the first microscopic field. This will give the first apoptosis score (A1 based on N1 cells). In the second field, the process is repeated and running scores recorded to give a running mean (A2 based on N2 cells). This process is repeated to give multiple running averages (A3, N3 . . . An, Nn). If these are plotted, the mean will be seen to wander and eventually oscillate about a mean value, and as N increases, this will become less. This procedure can then define experimentally the number of events to be assessed to produce a given quality of data.

4. Notes

1. Specificity: TUNEL/ISEL can by no means be said to be specific for apoptosis. Fortunately, the discrimination between large numbers of stained cells in an area of necrosis and the presence of scattered cells undergoing apoptosis should not pose a problem to a trained histologist. However, the effects of speed of fixation and the penetration rates of fixative will also influence the staining characteristics of the tissue. For instance, if a large piece of tissue is immersed in fixative,

there will be a significant delay before cells in the center of the tissue are fixed. For most purposes, TUNEL/ISEL can be considered as selective (rather than specific) for apoptosis in histological material. The technique assists with the identification and quantitation of apoptosis, but must be used in conjunction with simple morphological examination to exclude artifactual staining caused by technical aspects or staining as a result of demonstrating DNA strand breaks resulting from other physiological or pathological processes.

2. DAB is harmful and a possible carcinogen. Handle with care and deactivate solutions with bleach after use. Alternative peroxidase substrates with different colored products are available commercially (e.g., Dako, Vector Labs).

3. A number of kits are available commercially, for example, the *in situ* cell-death detection kits (Roche) or the FragEL kits (Calbiochem/ Oncogene Research). However, these kits are expensive and there is no reason why the component parts cannot be purchased from other competitive suppliers. The ability to undertake these methods and interpret the results is critically based on having good histology and immunostaining skills. If you do not have these, go to a laboratory that has them to learn.

4. For tissues or cells that are not paraffin embedded but have been fixed in paraformaldehyde or methanol/acetone, slides/coverslips may be air dried and then stored at –80°C individually wrapped in foil. Before staining, allow the wrapped slides to warm to room temperature before unwrapping to avoid condensation forming on the cells. Slides cannot be refrozen.

5. For cells or frozen sections fixed in paraformaldehyde, begin with the proteinase K treatment (**Subheading 3.2., step 4**). For cells or frozen sections fixed in methanol/acetone, begin the procedure at **step 6**.

6. The conditions for proteolytic digestion may need to be varied, and it is advisable in the first instance to perform a series of digestions, varying the concentration of proteinase K from 1–20 µg/mL.

7. To avoid contamination by DNases, solutions, tubes, and tips used in the TUNEL/ISEL procedure must be sterile.

8. Biotin-dUTP or digoxigenin-dUTP (Roche) can be substituted for the biotinylated dATP at the same concentration (but note that the unlabeled nucleotides should then be dATP, dCTP, and dGTP). For digoxigenin, the reaction is detected with antidigoxigenin antibodies (e.g., peroxidase conjugated antidigoxigenin, cat. no. 1207733,

diluted 1:200). Nucleotide triphosphates labeled with fluorescent markers are also available. These can be identified directly with fluorescence microscopy, or enzyme-linked anti-FITC antibodies can be used.

9. It is important to use the correct concentration of Klenow (or TdT) enzyme. Excess enzyme will lead to nonspecific staining of morphologically normal nuclei whereas insufficient enzyme will lead to a reduction in the staining of apoptotic nuclei *(21,22)*. We advise titrating the enzyme concentration in preliminary experiments to obtain the optimal conditions for your samples.

10. Modifications of the basic method are well documented for flow cytometry *(23)*, and kits are available commercially.

11. Controls: As with any technique, it is essential to perform a number of controls. A positive control should be included with each batch, to test for variations in the intensity of staining from day to day. For each sample, an appropriate immunohistochemical control should be performed in parallel, to identify staining caused by endogenous enzyme activity and/or nonspecific binding of the detection reagents. This is most easily achieved by exclusion of the enzyme or the labeled nucleotide from the TUNEL/ISEL reaction mixture.

12. This method uses peroxidase detection of biotinylated nucleotides with DAB and is highly suited to paraffin sections. The protocol should be amended if a different label has been used or if a different detection system is preferred (e.g., fluorescent detection).

13. Staining patterns and their interpretation: Irrespective of the enzyme or label employed, not all apoptotic bodies are intensely stained and it is not uncommon to see that the extremely condensed nuclei are relatively unstained. The nuclei of nonapoptotic cells should always be unstained. A generalized staining of all or many apparently normal nuclei suggests that proteolytic digestion has been too harsh—a similar effect is seen if a DNase treatment is used as a positive control *(21,22)*. Necrotic cells are also stained by the method *(28)*, and there are a number of situations in which staining can be seen in morphologically normal nuclei, for example, in spermatogonia *(22)* and after exposure to some DNA damaging agents *(44)*. DNA breaks could be present as the result of fixation and processing procedures, which result in the accumulation of lower molecular weight DNA. In addition, the action of section cutting and pretreatments such as exposure to hydrogen peroxide to block endogenous peroxidase activity might also cause DNA breaks in cells. In practice, this means that a

range of concentrations of protease and polymerase/TdT may need to be tested, particularly when using the technique for the first time.

References

1. Kroemer, G. (2003) Mitochondrial control of apoptosis: an introduction. *Biochem. Biophys. Res. Commun.* **304**, 433–435.
2. Lawen, A. (2003) Apoptosis—an introduction. *Bioessays* **25**, 888–896.
3. Reed, J. C. (2003) Apoptosis-targeted therapies for cancer. *Cancer Cell* **3**, 17–22.
4. Operfman, J. T. and Korsmeyer, S. J. (2003). Apoptosis in the development and maintenance of the immune system. *Nat. Immunol.* **4**, 410–415.
5. Kerr, J. F. R., Wyllie, A. H., and Currie, A. R. (1972) Apoptosis: a basic biological phenomenon with wide ranging implications in tissue kinetics. *Br. J. Cancer* **26**, 239–257.
6. Glucksman, A. (1951) Cell deaths in normal vertebrate ontogeny. *Biol. Rev.* **26**, 59–86.
7. Saunders, J. W. (1966) Death in embryonic systems. *Science* **154**, 604–612.
8. Sanderson, C. J. (1976) The mechanism of T-cell mediated cytotoxicity II. Morphological studies of cell death by time-lapse microcinematography. *Proc. Roy. Soc. Lond. B.* **192**, 241–255.
9. Matter, A. (1979) Microcinematographic and electron microscopic analysis of target cell lysis induced by cytotoxic T lymphocytes. *Immunology* **36**, 179–190.
10. Coles, H. S. R., Burne, J. F., and Raff, M. C. (1993) Large-scale normal cell death in the developing rat kidney and its reduction by epidermal growth factor. *Development* **118**, 777–784.
11. Hall, P. A., Coates, P. J., Ansari, B., and Hopwood, D. (1994) Regulation of cell number in the mammalian gastrointestinal tract: the importance of apoptosis. *J. Cell Sci.* **107**, 3569–3577.
12. Potten, C. S. (1996) What is an apoptotic index measuring? A commentary. *Br. J. Cancer* **74**, 1743–1748.
13. Wyllie, A. H., Kerr, J. F. R., and Currie, A. R. (1980) Cell death: the significance of apoptosis. *Int. Rev. Cytol.* **68**, 251–306.
14. Kerr, J. F. R., Searle, J., Harmon, B. V., and Bishop, C. J. (1987) Apoptosis, in *Perspectives on Mammalian Cell Death* (Potten, C. S., ed.), Oxford Science Publications, Oxford, UK, pp. 93–128.

15. Wyllie, A. H. (1980) Glucocorticoid-induced thymocyte apoptosis is associated with endogenous endonuclease activation. *Nature* **284,** 555–556.

16. Wyllie, A. H., Morris, R. G., Smith, A. L., and Dunlop, D. (1984) Chromatin cleavage in apoptosis: association with condensed chromatin morphology and dependence on macromolecular synthesis. *J. Pathol.* **142,** 67–77.

17. Zakeri, Z. F., Quaglino, D., Latham, T., and Lockshin, R. A. (1993) Delayed internucleosomal DNA fragmentation in programmed cell death. *FASEB J.* **7,** 470–478.

18. Walker, P. R., Smith, C., Youdale, T., Leblanc, J., Whitfield, J. F., and Sikorska, M. (1991) Topoisomerase II-reactive chemotherapeutic drugs induce apoptosis in thymocytes. *Cancer Res.* **51,** 1078–1085.

19. Fehsel. K., Kolb-Bachofen, V., and Kolb, H. (1991) Analysis of TNF alpha-induced DNA strand breaks at the single cell level. *Am. J. Pathol.* **139,** 251–254.

20. Gavrieli, Y., Sherman, Y., and Ben-Sasson, S. A. (1992) Identification of programmed cell death in situ via specific labelling of nuclear DNA fragmentation. *J. Cell Biol.* **119,** 493–501.

21. Wijsman, J. H., Jonker, R. R., Keijzer, R., van de Velde, C. J. H., Cornelisse, C. J., and van Dierendonck, J. H. (1993) A new method to detect apoptosis in paraffin sections: in situ end-labelling of fragmented DNA. *J. Histochem. Cytochem.* **41,** 7–12.

22. Ansari, B., Coates, P. J., Greenstein, B. D., and Hall, P. A. (1993) In situ end-labelling detects DNA strand breaks in apoptosis and other physiological and pathological states. *J. Pathol.* **170,** 1–8.

23. Gold, R., Schmied, M., Rothe, G., Zischler, H., Breitschopf, H., Wekerle, H., and Laussmann, H. (1993) Detection of DNA fragmentation in apoptosis: application of in situ nick translation to cell culture systems and tissue sections. *J. Histochem. Cytochem.* **41,** 1023–1030.

24. Mundle, S. D., Gao, X. Z., Khan, S., Gregory, S. A., Preisler, H. D., and Raza, A. (1995) Two in situ end labelling techniques reveal different patterns of DNA fragmentation during spontaneous apoptosis in vivo and induced apoptosis in vitro. *Anticancer Res.* **15,** 1895–1904.

25. Migheli, A., Attanasio, A., and Schiffer, D. (1995) Ultrastructural detection of DNA strand breaks in apoptotic neural cells by in situ end-labelling techniques. *J. Pathol.* **176,** 27–35.

26. Kurrer, M. O., Pakala, S. V., Hanson, H. L., and Katz, J. D. (1997) Beta cell apoptosis in T cell-mediated autoimmune diabetes. *Proc. Natl. Acad. Sci. USA* **94**, 213–218.

27. Strater, J., Walczak, H., Krammer, P. H., and Moller, P. (1996) Simultaneous in situ detection of mRNA and apoptotic cells by combined hybridization and TUNEL. *J. Histochem. Cytochem.* **44**, 1497–1499.

28. Coates, P. J. (1994) Molecular methods for the identification of apoptosis in tissues. *J. Histotechnol.* **17**, 261–267.

29. Coates, P. J., Hales, S. A., and Hall, P. A. (1996) The association between cell proliferation and apoptosis: studies using the cell cycle associated proteins Ki-67 and DNA polymerase alpha. *J. Pathol.* **178**, 71–77.

30. Didenko, V. V., Tunstead, J. R., and Hornsby, P. J. (1998) Biotin-labeled hairpin oligonucleotides: probes to detect double-strand breaks in DNA in apoptotic cells. *Am. J. Pathol.* **152**, 897–902.

31. Hornsby, P. J. and Didenko, V. V. (2002) In situ DNA ligation as a method for labeling apoptotic cells in tissue sections. An overview. *Methods Mol. Biol.* **203**, 133–141.

32. Didenko, V. V., Ngo, H., and Baskin, D. S. (2003) Early necrotic DNA degradation: presence of blunt-ended DNA breaks, 3' and 5' overhangs in apoptosis, but only 5' overhangs in early necrosis. *Am. J. Pathol.* **162**, 157–1578.

33. Marshman, E., Ottewell, P. D., Potten, C. S., and Watson, A. J. M. (2001) Caspase activation during spontaneous and radiation-induced apoptosis in the murine small intestine. *J. Pathol.* **195**, 285–292.

34. Duan, W. R., Garner, D. S., Williams, S. D., Funckes-Shippy, C. L., Spath, I. S., and Blomme E. A. G. (2003) Comparison of immunohistochemistry for activated caspase 3 and cleaved cytokeratin 18 with the TUNEL method for quantification of apoptosis in histological sections of PC-3 subcutaneous xenografts. *J. Pathol.* **199**, 221–228.

35. Carr, N. J. (2000) M30 expression demonstrates apoptotic cells, correlates with in situ end-labelling and is associated with Ki67 expression in large intestinal neoplasia. *Arch. Pathol. Lab. Med.* **124**, 1768–1772.

36. Woo, M., Hakem, R., Soengas, M. S., Duncan, G. S., Shahinian, A., et al. (1998) Essential contribution of caspase 3/CPP32 to apoptosis and its associated nuclear changes. *Genes Dev.* **12**, 806–819.

37. Liang, Y., Yan C., and Schor, N.F. (2001) Apoptosis in the absence of caspase 3. *Oncogene* **20**, 6570–6578.

38. Newton, K. and Strasser, A. (2003) Caspases signal not only apoptosis but also antigen-induced activation in cells of the immune system. *Genes Dev.* **17,** 819–825.

39. Frankfurt, O. S., Robb, J. A., Sugarbaker, E.V., and Villa, L. (1996). Monoclonal antibody to single-stranded DNA is a specific and sensitive cellular marker of apoptosis. *Exp. Cell Res.* **226,** 387–397.

40. Watanabe, I., Toyoda, M., Okuda, J., Tenjo, T., Tanaka, K., Yamamoto, T., et al. (1999). Detection of apoptotic cells in human colorectal cancer by two different in situ methods: antibody against single stranded DNA and terminal deoxynucleotidyl transferase-mediated dUTP-biotin nick end-labeling (TUNEL) methods. *Jpn. J. Cancer Res.* **90,** 188–193.

41. Garden, G. A., Bothwell, M., and Rubel, E. W. (1991) Lack of correspondence between mRNA expression for a putative cell death model (SGP-2) and neuronal cell death in the central nervous system. *J. Neurobiol.* **22,** 590–604.

42. Szondy, Z., Molnar, P., Nemes, Z., Boyiadzis, M, Kedei, N., Toth, R., et al. (1997) Differential expression of tissue transglutaminase during in vivo apoptosis of thymocytes induced via distinct signalling pathways. *FEBS Lett.* **404,** 307-313.

43. Hilton, D. A., Love, S., and Barber, R. (1997) Demonstration of apoptotic cells in tissue sections by in situ hybridization using digoxigenin-labelled poly(A) oligonucleotide probes to detect thymidine-rich DNA sequences. *J. Histochem. Cytochem.* **45,** 13.

44. Coates, P. J., Save, V., Ansari, B., and Hall, P. A. (1995) Demonstration of DNA damage/repair in individual cells using in situ end labelling: association of p53 with sites of DNA damage. *J. Pathol.* **176,** 19-26.

45. Aherne, W. A. and Dunnill, M. S. (1982) *Morphometry.* Edward Arnold, London, UK.

6

Immunoassay for Single-Stranded DNA in Apoptotic Cells

Oskar S. Frankfurt

Summary

The apoptosis assay described in this chapter is based on the selective denaturation of DNA in condensed chromatin of apoptotic cells and the detection of denatured DNA with a monoclonal antibody highly specific to single-stranded DNA. Optimal results are obtained by the heating at a relatively low temperature in the presence of formamide. The assay detects apoptotic cells but not necrotic cells or cells with DNA breaks in the absence of apoptosis. The sensitivity of the assay reflects the detection of early and late apoptosis. Apoptotic cells are detected in the sections of frozen or formalin-fixed paraffin-embedded tissues by immunohistochemistry and in the cell suspensions by flow cytometry or fluorescence microscopy. Apoptosis enzyme-linked immunoassay based on DNA denaturation by formamide in microtiter plates and one-step immunostaining is applied for high-throughput screening of drugs. The enzyme-linked immunoassay has the ability to distinguish anticancer drugs from toxic chemicals, to predict selective toxicity to cancer cells, and to detect drug synergism.

Key Words: Apoptosis; single-stranded DNA; ELISA; drug screening; monoclonal antibody; immunohistochemistry; flow cytometry.

1. Introduction

Specific and sensitive cellular markers are necessary for the detection and quantitative analysis of apoptosis. Identification of apoptotic cells by specific markers in histological sections is especially

From: *Methods in Molecular Biology, vol. 282: Apoptosis Methods and Protocols*
Edited by: H. J. M. Brady © Humana Press Inc., Totowa, NJ

important for heterogenous cell populations, such as occurs in normal and neoplastic tissues. Histochemical analysis of apoptosis in tissue sections is critical because morphological evaluation does not provide accurate counts of apoptotic cells, and biochemical analysis of DNA breaks gives no information about cell types undergoing apoptotic death.

In this chapter, a novel immunochemical method for the detection of apoptotic cells is described *(1–3)*. Most investigators at the present time rely on terminal deoxynucleotidyl transferase-mediated dUTP nick-end labeling (TUNEL) staining to detect apoptotic cells and to evaluate the role of apoptosis in disease. However, several studies demonstrated that TUNEL is not specific for apoptosis, because it also detects necrotic and autolytic types of cell death *(2,4)*. The sensitivity of TUNEL is compromised, because it detects only late stages of apoptosis associated with the low-mol-wt DNA fragmentation *(2)*. The application of TUNEL is also limited by the fact that in various cell types, apoptosis is not accompanied by internucleosomal DNA fragmentation and therefore is not detected by TUNEL. Therefore, a specific and sensitive cellular marker based on a different mechanism than TUNEL is needed to determine the role of apoptotic death in biology and pathology.

The method for the identification of apoptotic cells described here is based on the staining of cell suspensions and tissue sections with monoclonal antibodies (MAbs) to single-stranded DNA (ssDNA). The procedure includes three steps: fixation, heating, and staining with MAbs. The critical step is the heating of cells or sections, which is performed in conditions inducing DNA denaturation only in apoptotic cells. The selective thermal denaturation reflects decreased stability of DNA induced by the digestion of nuclear proteins during apoptosis. The fact that proteolysis is responsible for DNA denaturation was demonstrated by the elimination of staining in apoptotic cells reconstituted with histones and by the induction of staining in nonapoptotic cells treated with proteolytic enzymes *(3)*. The effect of proteolysis on MAb staining is in agreement with the ability of histones to stabilize DNA against thermal denaturation *(5)* and with the digestion of histones during apoptosis *(6)*.

The higher sensitivity of MAb staining compared to TUNEL reflects the different mechanisms of the two techniques. TUNEL detects low-mol-wt DNA fragmentation associated with late apoptosis, whereas MAbs to ssDNA detect the early stages of apoptosis and stain apoptotic cells in the absence of low-mol-wt DNA fragmentation *(2,3)*. These advantages of the MAb method are based on the fact that protease activation is an early and universal event in apoptosis *(6)*. Importantly, in contrast with the TUNEL method, MAbs to ssDNA are specific for apoptotic cell death and do not detect necrotic cells *(1,2)*.

Initially, MAbs to ssDNA were applied in our studies for the detection of DNA breaks induced by cytotoxic agents *(7,8)*. Heating of fixed cells suspended in phosphate-buffered saline containing a low concentration of Mg^{2+} induced DNA denaturation in cells treated with alkylating agents, but did not affect DNA conformation in untreated cells. There was a linear relation between MAb binding and the loss of cell viability *(9)*. The method proved to be useful for the analysis of DNA damage and repair in individual cells, for the detection of drug-resistant cell subsets, and made possible the discovery of intercellular transfer of drug resistance *(10)*.

The critical role of Mg^{2+} ions for the detection of DNA damage with anti-ssDNA MAbs was established in these studies *(7,8)*. Heating of cells suspended in medium without Mg^{2+} induced DNA denaturation and antibody binding in both treated and untreated cells. In the presence of 0.5–1.25 m*M* $MgCl_2$, only less stable DNA with drug-induced breaks was denatured. Higher concentrations of Mg^{2+} decreased MAb binding in drug-treated cells. The effects of Mg^{2+} on DNA denaturation in fixed cells is consistent with the stabilization of DNA in solution against thermal denaturation, which is achieved by the neutralization of negative charges in phosphate groups *(11)*.

Cell lines in which drug treatment did not induce apoptosis were used for the analysis of DNA breaks with MAbs to ssDNA *(9,10)*. Only after the technique was applied to chronic lymphocytic leukemia (CLL) cells was the MAb staining of apoptotic cells discovered *(12)*. A subset of cells having a decreased DNA content and exhibiting intense MAb fluorescence and typical apoptotic morphology was observed in cultures of CLL cells. Although MAb binding was significantly higher in apoptotic cells than in nonapoptotic cells with DNA breaks, it was important to develop conditions under which only DNA in apoptotic cells was denatured and stained with MAbs to ssDNA. The selective denaturation of DNA in apoptotic cells was achieved by increasing the $MgCl_2$ concentration during heating to 2.5–5 m*M (1,2)*.

It is important to note that MAbs F7-26 and AP-13 are specific for DNA in single-stranded conformation, and that the conditions of DNA denaturation determine the type of cellular damage detected by the procedure *(1–3,7–10)*. In environments that destabilize DNA (e.g., low ionic strength, acid treatment), normal DNA will be stained by these MAbs. Heating performed under conditions that moderately stabilize DNA (e.g., low Mg^{2+} concentration) will induce staining of DNA with breaks, while under conditions inducing maximal DNA stability (e.g., high concentration of Mg^{2+}), only DNA in apoptotic cells will denature and bind the antibody. MAb binding is not associated with DNA replication as demonstrated by the absence of staining of S phase cells *(1–3)*. Probably, the digestion of DNA-bound proteins, such as histones, in apoptotic cells induces a high level of DNA instability to thermal denaturation, which is not prevented with the neutralization of phosphate groups by Mg^{2+}. F7-26 and AP-13 differ with respect to antigenic determinant (deoxycytidine and thymidine, respectively) and the size of DNA in single-stranded conformation necessary for binding. F7-26 binds to smaller stretches of ssDNA, which may explain the shorter heating time and the lower temperature required to effect its binding to DNA in apoptotic cells.

The relation between the intensity of drug-induced cellular damage and the binding of the antibody to DNA is different in nonapoptotic and apoptotic cells. Antibody binding characterized by mean fluorescence intensity in the total cell population is proportional to the drug dose when DNA breaks are measured in nonapoptotic cells *(9)*. In contrast, fluorescence of the antibody is similar in apoptotic cells at various drug doses, and only the number of apoptotic cells is varied as a function of drug dose *(13,14)*. These observations are consistent with the notion that apoptosis is an all-or-none phenomenon, which once triggered induces a similar type of damage.

In conclusion, the immunoassay for ssDNA in apoptotic cells is a procedure based on the selective thermal denaturation of apoptotic DNA and the staining of cells with MAbs highly specific for DNA in single-stranded conformation. The low stability of apoptotic DNA to thermal denaturation is induced by the digestion of DNA-bound proteins during early stages of apoptosis and, in contrast to DNA instability induced by breaks, is not prevented by the presence of Mg^{2+} in the heating medium. MAbs to ssDNA provide a cellular marker specific for apoptotic death that is independent of internucleosomal DNA fragmentation and useful for the detection of different stages of apoptosis in various cell types. The high sensitivity of the assay reflects the central role of proteolysis in the initiation and execution of apoptosis. The procedure is outlined in **Fig. 1**.

2. Materials

2.1. Staining of Cell Suspensions for Flow Cytometry and Fluorescence Microscopy

1. Phosphate-buffered saline (PBS): Dubecco's PBS without $CaCl_2$ and without $MgCl_2$ (Gibco BRL): 0.2 g KCl, 0.2 g KH_2PO_4, 2.16 g Na_2HPO_4, 8 g NaCl, distilled H_2O to 1 L, pH 7.2.
2. Methanol, 100%, precooled to –20°C.
3. 63 m*M* Magnesium chloride solution: Dissolve 6 mg of $MgCl_2$ (anhydrous, Sigma) per mL of dH_2O. Prepare fresh.
4. MAb F7-26 specific for ssDNA. Working concentration: 10 μg/mL in PBS supplemented with 5% fetal bovine serum (FBS). Keep frozen at –20°C or –80°C. (APOSTAIN, Inc. [305]-868-3998; Fax: [305]-868-3445; E-mail: apostain@bellsouth.net; Website: www.apostain.com).
5. Fluorescein-conjugated goat antimouse IgM (Sigma): Working concentration: 1:50 in PBS supplemented with 5% FBS. Store frozen.
6. Propidium iodide, 1 μg/mL in PBS. Store at 4°C in the dark. Stable for 4–8 wk.

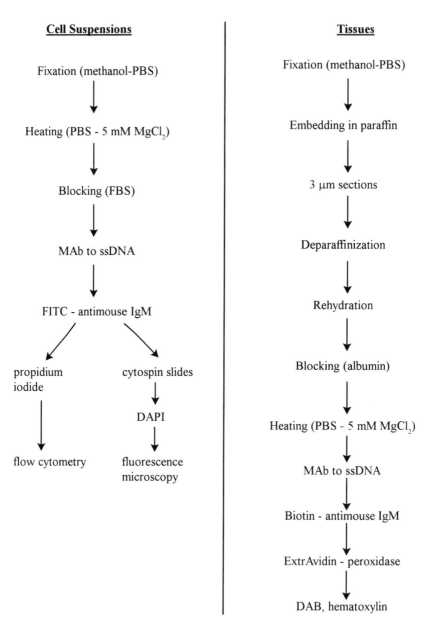

Fig. 1. Schematic diagram of the protocol for the staining of apoptotic cells with MAbs to ssDNA.

7. 4'-6-Diamidino-2-phenylindole (DAPI), 0.1 μg/mL, in PBS. Store at 4°C in the dark.
8. Saccomano cytology collection fluid (Baxter).
9. Vectashield mounting medium for fluorescence (Vector).
10. S1 nuclease (Sigma). Working concentration: 100 U/mL in acetate buffer (0.03 M sodium acetate, 1 mM ZnSO$_4$, pH 4.6). Store frozen at –20 or –80°C.
11. Histone solution: Type IIIS, lysine-rich fraction from calf thymus (Sigma), 0.25 mg/mL in PBS. Store frozen at –20°C.
12. Pyrex heavy-duty 15-mL centrifuge tubes or Kimble disposable 15-mL glass centrifuge tubes (Baxter).
13. Lauda circulating water bath M20 (Brinkman Instruments) or a hotplate.

2.2. Staining of Tissue Sections

1. Fixative: Methanol-PBS, 6:1. Store at–20°C.
2. Methanol, 100%.
3. Xylene.
4. Paraffin.
5. PBS (*see* **Subheading 2.1.**, **item 1**).
6. 63 mM MgCl$_2$ (*see* **Subheading 2.1.**, **item 3**).
7. 10% Triton X-100.
8. SafeClear tissue clearing agent (Curtis-Matheson).
9. Conical 50-mL polypropylene centrifuge tubes (Sarstedt).
10. 3% Hydrogen peroxide solution.
11. Bovine serum albumin (BSA): 0.1% in PBS.
12. MAb F7-26 specific to ssDNA (*see* **Subheading 2.1.**, **item 4**).
13. Biotin-conjugated rat monoclonal antimouse IgM (Zymed): Working concentration: 1:50 in PBS supplemented with 0.2% Tween-20 and 0.1%, sodium azide. Store at 4°C. Stable for 2–3 mo. **Caution: Sodium azide is highly toxic.**
14. ExtrAvidin-peroxidase (Sigma): Stock solution 100 μg/mL in PBS; store at –20°C. Working concentration: 10 μg/mL, prepare fresh.
15. Liquid DAB-plus substrate kit (Zymed).
16. Lerner Hematoxylin.
17. Mounting medium (Baxter).
18. Water bath (*see* **Subheading 2.1.**, **item 13**).

3. Methods

3.1. Detecting ssDNA in Cell Suspensions

3.1.1. Fixation (see **Note 1**)

1. Centrifuge 1–2×10^7 cells at $200g$ for 5 min.
2. Decant, and resuspend the pellet in 1 mL of PBS.
3. Slowly add 6 mL of cold methanol while vortexing.
4. Store the fixed cells at $-20°C$ for 16–24 h before staining.

3.1.2. Heating (see **Notes 2–5**)

1. Distribute 0.5–1.0×10^6 fixed cells into glass tubes, centrifuge, and decant the fixative.
2. Resuspend the pellet in 0.4 mL of PBS freshly supplemented with 5 mM $MgCl_2$ (9.2 mL of PBS + 0.8 mL of $MgCl_2$).
3. Immerse the rack with tubes into a circulating water bath (preheated to 99°C) for 5 min or into a beaker with boiling water on a hotplate for 5 min.
4. Place the rack in ice-cold water for 10 min.

3.1.3. Blocking (see **Note 6**)

Add 0.4 mL of 40% FBS in PBS and incubate on ice for 15 min.

3.1.4. Staining (see **Notes 6** and **7**)

1. Centrifuge the cells at $200g$ for 5 min.
2. Resuspend the pellet in 100 µL of MAb F7-26 solution.
3. Incubate at room temperature for 30 min.
4. Rinse twice in PBS.
5. Resuspend the pellet in 100 µL of FITC-conjugated antimouse IgM.
6. Incubate at room temperature for 30 min.
7. Rinse once in PBS.
8. For flow cytometry: Resuspend the pellet in 0.5 mL of propidium iodide solution.
9. For fluorescence microscopy: Resuspend the pellet in Saccomano fluid, stain cytospin slides with DAPI for 10 min, rinse with PBS, dry, and mount in VectaShield.

3.1.5. Analysis

1. Flow cytometry measurements are performed using log scale for green fluorescence from fluorescein-labeled antibody and linear scale for DNA-bound propidium iodide. For example, FL 1 and FL 2 settings for FACScan were 440/log and 400/1.85, respectively, for etoposide-treated MOLT-4 cells. Typical fluorescence contour plots are presented in ref. *(1,2)*.
2. Slides are observed in a fluorescence microscope using UV excitation for the DNA-fluorochrome DAPI and 450–490 nm excitation for fluorescein-labeled antibody. Dual-labeling makes it possible to characterize chromatin distribution in positive cells by changing excitation filters *(1,2)*.

3.1.6. Controls (see **Notes 8–10**)

The following two controls are recommended:

1. Incubate the cells in histone solution for 30 min at room temperature before heating, that is, before **step 2** of **Subheading 3.1.2.** Reconstitution of apoptotic nuclei with lysine-rich histones completely suppresses MAb binding.
2. After heating, treat the cells with S1 nuclease at 37°C for 30 min. Digestion of ssDNA eliminates MAb binding. Buffer alone has no effect on the staining.

3.2. Detecting ssDNA in Tissue Sections

3.2.1. Fixation and Embedding (see **Notes 1, 11**, and **12**)

1. Fix fresh tissue in methanol-PBS at –20°C for 1–3 d.
2. Dehydrate the fixed tissue in two changes of absolute methanol (1 h each) and two changes of xylene (1 h each).
3. Incubate in two changes of paraffin at 56°C (1 h each).
4. Embed in paraffin.
5. Cut 3-µm sections from paraffin blocks.
6. Attach sections to superfrost/plus slides, and heat at 56°C for 1–2 h.

3.2.2. Deparafinization and Rehydration (see **Notes 13** and **14**)

1. Incubate the slides in two changes of SafeClear (15 min each).
2. Incubate the slides in three changes of methanol-PBS (20 min each).

3. Rinse with PBS.
4. Incubate the slides in PBS supplemented with 0.2% Triton X-100 and 5 m*M* MgCl$_2$ for 5 min.

3.2.3. Heating (see **Notes 2–5**)

1. Transfer the slides into 50-mL centrifuge tubes containing 30 mL of room temperature PBS freshly supplemented with 5 m*M* MgCl$_2$ (27.6 mL PBS + 2.4 mL MgCl$_2$).
2. Immerse the rack containing the centrifuge tubes into a circulating water bath (preheated to 99°C) for 5 min or into a beaker of boiling water on a hotplate for 5 min.
3. Remove the slides with forceps from the centrifuge tubes, and transfer to tubes containing ice- cold PBS for 10 min.

3.2.4. Staining (see **Notes 6** and **7**)

1. Incubate the slides in 3% H$_2$O$_2$ for 5 min to block endogenous peroxidase activity.
2. Rinse twice with PBS.
3. Treat the slides with 0. 1% BSA at room temperature for 30 min.
4. Rinse twice with PBS.
5. Apply anti-ssDNA MAb F7-26 to the top of the tissue section (100 μL/slide).
6. Incubate at room temperature for 15 min. Rinse twice with PBS.
7. Apply biotin-conjugated rat antimouse IgM for 15 min, and then rinse twice with PBS.
8. Apply ExtrAvidin-peroxidase for 15 min, and then rinse with PBS.
9. Apply chromogen solution (DAB), counterstain with hematoxylin, dehydrate, and mount.

3.2.5. Controls (see **Notes 8–10**)

The following four controls are recommended:

1. Following deparafinization (i.e., **steps 1–3** of **Subheading 3.2.2.**), treat the tissue sections with proteinase K (2 μg/mL in PBS) at 37°C for 20 min, rinse with PBS and proceed to **step 4** of **Subheading 3.2.2.** All nuclei are stained (positive control).

2. Heat the slides immersed in dH_2O rather than in $PBS/MgCl_2$ solution. All nuclei are brightly stained (positive control).
3. Following deparafinization, incubate the sections in histone solution for 20 min, rinse with PBS, and proceed to **step 4** of **Subheading 3.2.2.** Staining of apoptotic nuclei is eliminated (negative control).
4. Rinse the heated sections with saline, treat with S 1 nuclease at 37°C for 20 min, rinse with PBS, and proceed to **step 1** of **Subheading 3.2.4.** Staining of apoptotic nuclei is eliminated (negative control).

4. Assay Based on Formamide-Induced DNA Denaturation

We have recently modified the apoptosis immunoassay, expanded its applications to formalin-fixed tissues, and developed a solid-phase enzyme-linked immunoassay (ELISA) for high-throughput screening of apoptosis-inducing drugs. These applications are based on the selective denaturation of apoptotic DNA in the presence of formamide at a relatively low temperature, instead of denaturation in PBS at 100°C.

Formamide induces DNA denaturation in the apoptotic nuclei but has no such effect on DNA in the nonapoptotic cells *(15)*. Formamide-induced DNA denaturation combined with the detection of denatured DNA by the anti-ssDNA MAb made it possible to identify the apoptotic cells by immunohistochemistry, flow cytometry, and fluorescence microscopy. This procedure produced intense staining of the condensed chromatin in the apoptotic nuclei. In contrast, necrotic cells in cultures treated with sodium azide, saponin, or hyperthermia did not bind this antibody, demonstrating the specificity of the formamide–MAb assay for the apoptotic cells. TUNEL, however, stained 90–100% of necrotic cells in all three models of necrosis. The MAb did not stain cells with single- or double-stranded DNA breaks in the absence of apoptosis, indicating that staining of the apoptotic nuclei is induced by specific changes in condensed chromatin. Importantly, formamide-MAb technique identified apoptotic cells in frozen sections and in histological sections of formalin-fixed paraffin-embedded tissues. For detailed protocols, visit **www.apostain.com**.

Groos et al. *(16)* evaluated different techniques for the *in situ* detection of apoptosis in human and rat small intestinal epithelium and concluded that the antibody detecting formamide-denatured ssDNA in apoptotic cells was both suitable and reliable. In comparison with this, the TUNEL assay was less reliable and had to be adjusted for each specimen on the basis of the findings produced by other techniques. The adaptation of formamide–MAb technique for the detection of apoptotic cells in microtiter plates is described in ref. *17*. The apoptosis ELISA assay involves growth of cells in 96-well plates, treatment of the attached cells with formamide to denature DNA in apoptotic cells, and one-step immunostaining of the denatured DNA with a mixture of anti-ssDNA MAb and peroxidase-conjugated anti-mouse IgM. A near linear increase in absorbance was seen as the number of apoptotic cells per well increased from 500 to 5000. Untreated and necrotic cells or cells with single-stranded DNA breaks induced by H_2O_2 did not produce signal above the background level. In leukemic cell cultures treated with IC_{50} concentration of etoposide intense ELISA signal was detected. The ratio of absorbance values from drug-resistant and drug-sensitive cell lines treated with etoposide was in agreement with the degree of resistance determined by growth inhibition assays. These data demonstrated that formamide-MAb apoptosis ELISA provides sensitive and specific detection of apoptotic cells and may be used for high-throughput screening of drugs based on their ability to induce or suppress apoptosis. For ELISA protocols and the application of a one-step immunostaining kit, visit **www.apostain.com**.

The utility of apoptosis ELISA for the screening of chemical libraries was evaluated by the comparison of apoptosis-inducing activity of 13 clinically useful anticancer drugs and 12 toxic compounds without antitumor activity *(18)*. Two types of compounds were clearly distinguished by the ability to induce apoptosis in cultures of leukemic cells. The anticancer drugs induced intense ELISA reaction (mean absorbance 2.0) in cultures treated with two-fold of concentrations producing 50% inhibition of cell growth. In contrast to anticancer drugs, toxic chemicals did not increase apoptosis ELISA absorbance at cytotoxic doses. The application of apoptosis ELISA to chemosensitivity testing was also demonstrated by its ability to rapidly detect synergism of anticancer drug combinations. The sensitivity of apoptosis ELISA was similar to the sensitivity of long-term cell survival assayand significantly higher than that of the MTT assay.

The application of apoptosis ELISA for the detection of selective toxicity to cancer cells is described in ref. *19*. The effects of anticancer drugs and toxic compounds were evaluated in cultures of human breast cancer cells and normal diploid fibroblasts. The normal:cancer cells ratio of apoptosis inducing concentrations was in the range of 33–200 for anticancer drugs and 1.3–3.0 for toxins. The normal:cancer cells ratio of growth inhibiting concentrations was in a similar range for anticancer drugs and toxins. Anticancer drugs induced apoptosis in breast cancer cells at 0.0015–0.5 μM, whereas toxins were effective at much higher concentrations of 8.0–50.0 μM. Moreover, most toxins did not induce apoptosis in breast cancer cells. These data demonstrated that apoptosis assay but not MTT and SRB growth inhibition assays could predict selective toxicity to cancer cells. Apoptosis ELISA could be used to detect possible anticancer leads during drug screening by determining apoptosis induction in cancer cells or through a comparison of apoptosis inducing concentrations in normal and cancer cells.

5. Notes

1. Fixation in methanol-PBS produces optimal results. The fixative should be cooled to −20°C before addition to cells and tissues. Fixed material should be kept in freezer. Fixation of tissues at room temperature or in refrigerator must be avoided.
2. PBS supplemented with $MgCl_2$ should be prepared shortly before heating by mixing the stock solution of $MgCl_2$ with PBS. PBS supplied by Gibco BRL is recommended at least at the initial stage of application. $MgCl_2$ should be kept anhydrous, because the concentration of Mg^{2+} is critical for the specific staining.
3. Clean glass centrifuge tubes should be used for the heating of cell suspensions. The types of tubes should not be changed because the thickness of glass affects the process of heating. For the heating of slides with tissue sections, disposable polypropylene centrifuge tubes

can be used. The temperature of the PBS/MgCl$_2$ solution inside the tube with slides after 5 min of heating was found to be 8–9°C lower than the temperature in the water bath. The heating regimens described here were selected for MAb F7-26 and for the specific conditions (type of tubes, volume of fluid) indicated in the protocol. Although the exact time and temperature may have to be determined for the particular experimental conditions, the heating should be in the range of 99–100°C for 5 min.

4. A heating, circulating water bath with electronic temperature control and digital display is recommended, especially when large numbers of tubes are heated. Heating also may be performed by immersion of a rack with a small number of tubes into a beaker or vessel containing boiling water on a hotplate, or in a noncirculating water bath with boiling water. Tubes with cells or slides should always be kept in a rack during heating. In general, specific staining will be absent after insufficient heating, whereas excessive heating will induce DNA denaturation in nonapoptotic cells and produce nonspecific background staining.

5. For the optimal staining of cell suspensions, carefully remove the fixative, add PBS/MgCl$_2$, and heat the tubes as soon as possible. Rinsing of fixed cells in PBS before heating and delay between the addition of PBS/MgCl$_2$ and heating may decrease specific staining. The volume of fluid in which the cells are suspended during heating and the number of cells should be kept constant for reproducible results.

6. Blocking and an optimal concentration of the MAb to ssDNA are needed to obtain specific staining of apoptotic cells. Nonspecific binding of MAb F7-26 to methanol-fixed cells is blocked by FBS, whereas BSA is the best blocking agent for tissue. Bright staining of apoptotic cells and the absence of antibody binding to nonapoptotic cells is obtained with the recommended range of MAb F7-26 concentrations (*1–3*). Excessive concentration of the antibody may induce some nonspecific binding to nonapoptotic cells, although at all concentrations, the staining of apoptotic cells will be more intense.

7. Second-step reagents should not bind to methanol-fixed cells or tissues not treated with MAbs to ssDNA. Fluorescein-labeled antimouse IgM in PBS containing FBS or newborn calf serum is recommended for the staining of cell suspensions. The concentration and the type of serum that are needed to suppress the nonspecific binding of the antimouse antibody to methanol-fixed cell suspensions may vary,

depending on the cell and antibody type. Biotin-labeled rat monoclonal antimouse IgM and ExtrAvidin-peroxidase are optimal second-step reagents for paraffin sections of methanol-fixed tissues.

8. The following positive controls are recommended to determine the sensitivity of MAb staining. Treatment of cells and tissue sections with a proteolytic enzyme before heating in $MgCl_2$-supplemented PBS should induce staining of nonapoptotic cells. This procedure reproduces DNA instability induced by the digestion of nuclear proteins during apoptosis. Heating of cells or sections suspended in dH_2O should induce bright staining of all nonapoptotic nuclei, indicating that the procedure is adequate for the detection of denatured DNA. Cell suspensions from nontreated cultures or sections of tissues with low apoptotic indices are used for the positive controls. Cells in crypts of small intestine that are negative after standard staining should be positive in proteinase- treated or dH_2O-heated sections.

9. Two procedures are recommended to determine that the staining of apoptotic cells reflects the exposure of single-stranded regions in DNA destabilized by the digestion of DNA-bound proteins. Elimination of staining by S 1 nuclease demonstrates that only ssDNA is detected by MAb binding. Reconstitution of cells or sections with lysine-rich histones restores DNA stability in apoptotic cells and prevents DNA denaturation during heating. Cells or sections with high apoptotic indices should be used for the S 1 nuclease and histone negative controls.

10. The following experimental models are recommended to evaluate the staining of apoptotic cells with MAbs to ssDNA:

    a. Exponentially growing MOLT-4 cultures treated with 5 μM etoposide for 6 h.

    b. Monolayer cultures of MDA-MB-468 breast cancer cells treated with 0.5 μM staurosporine for 2–4 h or 15 μM cisplatin for 18 h. Floating cells with apoptotic morphology are stained by the MAbs, whereas 10–20% of attached cells at early stages of apoptosis are positive.

    c. Small intestine from untreated or hydroxyurea-treated mice (500 mg/ kg 4 h). Surface villous cells are positive in control tissue, but crypt cells with condensed and fragmented chromatin are stained in drug-treated mice.

    d. Mouse thymus removed 5 h after the injection of 100 mg/kg methylprednisolone.

11. The thickness of tissue specimens should not be more than 3–5 mm, because sections from poorly fixed tissues will not be stained. Incubation in xylene may be longer for larger specimens and should be continued until the tissue becomes clear.
12. Freshly cut sections should be used for best results. Background nonspecific staining may develop in sections after prolonged storage.
13. Tissue sections and cells should be kept moist at all steps of the procedure. Air-dried sections, cryostat sections, smears, and cytospin preparations are not suitable for staining with anti-ssDNA MAbs, because drying will prevent selective denaturation of DNA in apoptotic cells.
14. The use of SafeClear as a substitute for xylene for the deparaffinization of tissue sections is recommended. SafeClear is nontoxic and produces better results than xylene. Prolonged treatment of sections in methanol/PBS is needed to remove SafeClear before rehydration in PBS.

References

1. Frankfurt, O. S. (1994) Detection of apoptosis in leukemic and breast cancer cells with monoclonal antibody to single-stranded DNA. *Anticancer Res.* **14,** 1861–1870.
2. Frankfurt, O. S., Robb, J. A., Sugarbaker, E. V., and Villa, L. (1996) Monoclonal antibody to single-stranded DNA is a specific and sensitive cellular marker of apoptosis. *Exp. Cell Res.* **226,** 387–397.
3. Frankfurt, O. S., Robb, J. A., Sugarbaker, E. V., and Villa, L. (1997) Apoptosis in breast carcinomas detected with monoclonal antibody to single-stranded DNA: Relation to bcl-2 expression, hormone receptors, and lymph node metastases. *Clin. Cancer Res.* **3,** 465–471.
4. Grasl-Kraupp, B., Ruttkay-Nedecky, B., Koudelka, M., Burowska, K., Bursch, W., and Schulte-Herman, R. (1995) In situ detection of fragmented DNA (TUNEL assay) fails to discriminate among apoptosis, necrosis, and autolytic cell death: A cautionary note. *Hepatology* **21,** 1465–1468.
5. Tsai, Y. H., Ansevin, A. T., and Hnilica, L. S. (1975) Association of tissue-specific histones with deoxyribonucleic acid. Thermal denaturation of native, partially dehistonized, and reconstituted chromatins. *Biochemistry* **14,** 1257–1265.
6. Martin, S. J. and Green, D. R. (1995) Protease activation during apoptosis: death by a thousand cuts? *Cell* **82,** 349–352.

7. Frankfurt, O. S. (1987) Detection of DNA damage in individual cells by flow cytometric analysis using anti-DNA monoclonal antibody. *Exp. Cell Res.* **170**, 369–380.

8. Frankfurt, O. S. (1990) Decreased stability of DNA in cells treated with alkylating agents. *Exp. Cell Res.* **191**, 181–185.

9. Frankfurt, O. S., Seckinger, D., and Sugarbaker, E. V. (1990) Flow cytometric analysis of DNA damage and repair in the cells resistant to alkylating agents. *Cancer Res.* **50**, 4453–4457.

10. Frankfurt, O. S., Seckinger, D., and Sugarbaker, E. V. (1991) Intercellular transfer of drug resistance. *Cancer Res.* **51**, 190–1195.

11. Eichhorn, G. L. (1962) Metal ions as stabilizers of the deoxyribonucleic acid structure. *Nature* **194**, 474–475.

12. Frankfurt, O. S., Byrnes, J. J., Seckinger, D., and Sugarbaker, E. V. (1993) Apoptosis (programmed cell death) and the evaluation of chemosensitivity in chronic lymphocytic leukemia and lymphoma. *Oncol. Res.* **5**, 37–42.

13. Frankfurt, O. S., Seckinger, D., and Sugarbaker, E. V. (1994) Pleotropic drug resistance and survival advantage in leukemic cells with diminished apoptotic response. *Int. J. Cancer* **59**, 217–224.

14. Frankfurt, O. S., Seckinger, D., and Sugarbaker, E. V. (1994) Apoptosis and growth inhibition in sensitive and resistant leukemic cells treated with anticancer drugs. *Int. J. Oncol.* **4**, 481–489.

15. Frankfurt, O. S. and Krishan, A. (2001) Identification of apoptotic cells by formamide-induced DNA denaturation in condensed chromatin. *J. Histochem. Cytochem.* **49**, 369–378.

16. Groos, S., Reale , E., and Luciano, L (2003) General suitability of techniques for in situ detection of apoptosis in small intestinal epithelium. *Anat. Rec.* **272A**, 503–513.

17. Frankfurt, O. S. and Krishan, A. (2001) Enzyme-linked immunosorbent assay (ELISA) for the specific detection of apoptotic cells and its application to rapid drug screening. *J. Immunol. Methods* **253**, 133–144.

18. Frankfurt, O. S. and Krishan, A. (2003) Apoptosis enzyme-linked immunosorbent assay distinguishes anticancer drugs from toxic chemicals and predicts drug synergism. *Chem-Biol. Interact.* **145**, 89–99.

19. Frankfurt, O. S. and Krishan, A. (2003) Apoptosis-based drug screening and detection of selective toxicity to cancer cells. *Anti-Cancer Drugs* **14**, 555–561.

7

Methods to Measure Membrane Potential and Permeability Transition in the Mitochondria During Apoptosis

Naoufal Zamzami and Guido Kroemer

Summary

Mitochondrial membrane permeabilization (MMP) constitutes an early event of the apoptotic process. MMP affects both mitochondrial membranes. Inner MMP leads to the dissipation of the inner transmembrane potential and outer MMP culminates in the efflux of apoptogenic factors. The exact molecular mechanisms of MMP are still controversial. A growing body of data suggests that the cell death regulatory activity of Bcl-2 family members depends, at least in some instances, on their ability to modulate the opening of the mitochondrial permeability transition pore complex. Here, we will detail some experimental protocols designed to measure mitochondrial membrane potential and permeability transition, either in intact cells or in isolated mitochondria.

Key Words: Apoptosis; mitochondria; permeability transition.

1. Introduction

Programmed cell death or apoptosis is an essential physiological process required for a normal development and for maintenance of tissue homeostasis. It is now clear that mitochondria play a crucial role in apoptosis, and they appear to be in the heart of the cellular "decision" to die. Changes in the mitochondrial membrane permeability are early events during apoptosis. These changes have two

From: *Methods in Molecular Biology, vol. 282: Apoptosis Methods and Protocols*
Edited by: H. J. M. Brady © Humana Press Inc., Totowa, NJ

main consequences; namely, a first decrease in the mitochondrial inner membrane potential and the release of the mitochondrial inter-membrane factors through the outer membrane. After a variety of death stimuli, mitochondria are recruited into the apoptotic pathway by proapoptotic proteins from the Bcl-2 family (BAX, BAK, and BID). Proapoptotic Bcl-2 family members localize primarily to separate subcellular compartments in the absence of a death signal. Antiapoptotic members are often constitutively targeted to intracellular membrane organites, such as mitochondria, endoplasmic reticulum, or the nuclear membrane *(1–3)*. After a death signal, several proapoptotic members, including BAX, BAK, and BID, undergo a conformational change that enables them to target and integrate into mitochondrial membranes *(4–8)*. Cumulating data suggest that the cell death regulatory activity of BCL-2 family members depends on their ability to modulate mitochondrial function by regulating, at least in some instances, the opening of the mitochondrial permeability transition (PT) pore complex. The PT pore, also called mitochondrial megachannel or multiple conductance channel *(9–11)*, is a dynamic multiprotein complex located at the contact site between the inner and the outer mitochondrial membranes, one of the critical sites of metabolic coordination between the cytosol, the mitochondrial intermembrane space, and the matrix. The PT pore participates in the regulation of matrix Ca^{2+}, pH, and volume. It also functions as a Ca^{2+}-, voltage-, pH-, and redox-gated channel with several levels of conductance and little if any ion selectivity *(9–14)*. In isolated mitochondria, opening of the PT pore causes matrix swelling with consequent distension and local disruption of the mitochondrial outer membrane (whose surface is smaller than that of the inner membrane), release of soluble products from the intermembrane space, dissipation of the mitochondrial inner transmembrane potential ($\Delta\Psi_m$), and release of small molecules up to 1500 Daltons from the matrix through the inner membrane *(9,10, 15,16)*. Similar changes are found in apoptotic cells, perhaps with the exception of matrix swelling, which is only observed in a transient fashion, before cells shrink *(17–21)*. In several models of apoptosis, pharmacological inhibition of the PT pore is cytoprotective, suggesting that opening of the PT pore can be rate-limiting for the death process *(21–23)*. Moreover, the cytoprotective oncoprotein Bcl-2 has been shown to function as an endogenous inhibitor of the PT pore *(23–25)*. Some authors have suggested that other mechanisms not involving the PT pore may account for mitochondrial membrane permeabilization during apoptosis *(26,27)*. Irrespective of this possibility, mitochondrial membrane permeabilization is a general feature of apoptosis. Here, we will detail experimental procedures designed to measure mitochondrial membrane potential and permeability transition, either in intact cells or in isolated mitochondria.

2. Materials

2.1. Cell-Permeant Probes for Mitochondria

See **Table 1** for the details of properties and stock solutions of the probes used in the procedures described.

2.2. Equipment

1. Thomas potter with Teflon inlet.
2. Hamilton 500-µL syringe.
3. Corex 15-mL tubes.
4. Cylinder made of cork (*see* **Note 1**).
5. Centrifuge.

2.3. Mitochondrial Purification

1. Homogenization (H) buffer: 300 m*M* saccharose, 5 m*M* *N-tris* (hydroxymethyl)methyl-2-aminoethanesulfonic acid (TES), and 200 µ*M* ethylenebis(oxyethylenenitrilo)tetraacetic acid (EGTA), pH 7.2, stored at 4°C.
2. Percoll (P) buffer: 300 m*M* saccharose, 10 m*M* *N-tris*(hydroxymethyl) methyl-2-aminoethanesulfonic acid (TES), and 200 µ*M* EGTA, pH 6.9, stored at 4°C.
3. Solution A: 90 mL of P buffer plus 2.05 g of saccharose.
4. Solution B: 90 mL of P buffer plus 3.96 g of saccharose.
5. Solution C: 90 mL of P buffer plus 13.86 g of saccharose.
6. Percoll stock solution (100%, Pharmacia).

Table 1
Details of Properties and Stock Solutions of the Probes Used in the Procedures Described[a]

Probe	Fluorescence	Stock solution	Staining solution (in PBS)
DIOC$_6$ (3) 3,3' dihexyloxacarbocyanine iodide	Green fluorescence	40 μM (Ethanol)	800 nM
Mitotracker Red - CMXROS Chloromethyl- X-rosamine	Red fluorescence	1 mM (DMSO)	4 μM
TMRM Tetramethylrhodamine methyl ester	Orange fluorescence	1 mM (DMSO)	4 μM
JC-1 (5,5',6,6'-tetrachloro-1,1'3,3'tetra methyl benzimidazolylcarbocyanine iodide)	Green and Red fluorescence (dual-emission spectrum)	0.76 mM (DMSO)	40 μM
HE Dihydroethidine	Red fluorescence	10 mM (DMSO)	10 μM
Rh 123 Rhodamine 123	Green fluorescence	10 mM (Ethanol)	1 mM
NAO Nonyl Acridine Orange	Green fluorescence	10 mM (Ethanol)	10 μM

[a]All fluorescent probes should be protected from light and kept frozen at –20°C.

7. 18% Percoll: 6.55 mL of solution A + 1.45 mL of Percoll solution (freshly prepared).
8. 30% Percoll: 5.6 mL of solution B + 2.4 mL of Percoll solution (freshly prepared).
9. 60% Percoll: 3.2 mL of solution C + 4.8 mL of Percoll solution (freshly prepared).

2.4. Determination of Transmembrane Potential in Isolated Mitochondria

1. M buffer: 220 mM sucrose, 68 mM mannitol, 10 mM KCl, 5 mM KH$_2$PO$_4$, 2 mM MgCl$_2$, 500 µM EGTA, 5 mM succinate, 2 µM rotenone, 10 mM N-hydroxyethylpiperazine-N'-2-ethanesulfonate, pH 7.2
2. Mitochondria are purified as described in **Subheading 3.3.**, kept on ice for a maximum of 4 h, and resuspended in M buffer (or similar buffers).
3. CMXROS stock solution is prepared as described above. Rhodamine123 stock solution (10 mM in ethanol to be kept at –20°C, protected against light).
4. Advanced cytofluorometer capable of detecting isolated mitochondria.
5. Spectrofluorometer.

2.5. Swelling Assay

1. Instrument for absorbance analysis (Spectrophotometer or spectrofluorimeter).
2. Freshly prepared mitochondria.
3. Swelling buffer (SB; *see* **Note 2**): 0.2 M sucrose, 10 mM Tris-MOPS, pH 7.4, 5 mM succinate, 1 mM phosphate inorganic, 2 µM rotenone, 10 µM Tris-EGTA, pH 7.4.

3. Methods

3.1. Monitoring Mitochondrial Parameters by Flow Cytometry

1. Collect (0.3 to 0.5 mL of 1 × 10^5 cells in culture medium or 1× phosphate-buffered saline [PBS]) and keep them on ice until the staining. For the labeling of specific cellular subpopulations, *see* **Note 3**.
2. Add the following amounts of staining solutions to 1 mL of cell suspension: 25 µL of DiOC$_6$(3) (final concentration: 20 nM) or 25 µL of

CMXRos (final concentration: 100 nM) or 25 μL of JC-1 (final concentration 1 μM) and transfer tubes to a water bath kept at 37°C (*see* **Note 4**). After 15–20 min of incubation, return cells to ice. Do not wash cells. Include a control (*see* **Note 5**).

3. Perform cytofluorimetric analysis within 10 min (*see* **Note 6**).

3.2. Multiple Staining of Mitochondrial Parameters in Intact Cells

As a result of mitochondrial membrane permeabilization, cells hyperproduce reactive oxygen species, which in turn oxidize mitochondrial inner membrane cardiolipins. These alterations can be detected simultaneously using fluorescence-activated cell-sorting techniques. Therefore, in addition to the measurement of mitochondrial inner membrane potential, it is possible to determine the production of reactive oxygen species using dihydroethidine probe (*see* **Note 7**). Alternatively, the damage produced by reactive oxygen species in mitochondria can be determined via assessing the oxidation state of cardiolipin, a molecule restricted to the inner mitochondrial membrane. Nonyl acridine orange (NAO) interacts stoichiometrically with intact, nonoxidized cardiolipin *(28)*. By consequence, a reduction in NAO fluorescence indicates a decrease in cardiolipin content.

1. Cells (0.3 to 0.5 mL of 1×10^5 cells in culture medium or PBS) should be kept on ice until the staining. Before dihydroethidine (HE) staining, cells may be labeled with specific antibodies conjugated to fluorescein isothiocyanate . Cells can also be stained for both mitochondrial potential and reactive oxygen species production. Because ethidium emits red fluorescence, the probe for potential determination should emit green fluorescence [e.g., DiOC$_6$(3)].

2. For staining, add the following amounts of staining solutions to 1 mL of cell suspension: 25 μL of HE (final concentration: 2.5 μM), 10 μL of NAO (final concentration: 100 nm), or 25 μL of JC-1 (final concentration: 1 μM) and transfer tubes to a water bath kept at 37°C. After 15–20 min of incubation, return cells to ice. Do not wash cells.

3. Perform cytofluorimetric analysis within 10 min while gating forward and sideward scatters on viable, normal-sized cells (*see* **Note 8**).

3.3. Mitochondrial Purification From Mouse Liver

1. After euthanasia, the liver is rapidly removed, rinsed with cold H buffer, and cut into small pieces using a pair of scissors. All steps are performed at 4°C or on wet ice.
2. Homogenize tissue in the Potter-Thomas homogenizer (approx 20 strokes, approx 500 revolutions per min).
3. Distribute homogenate in two Corex tubes and centrifuge 10 min at 760g.
4. Recover supernatant. Resuspend pellet in H buffer, centrifuge as above, and recover supernatant. Join the two supernatants.
5. Centrifuge for 10 min at 8740g. In the meantime, prepare Percoll gradient. Transfer 4 mL of 60% Percoll solution into a Corex glass tube, then introduce the cork cylinder, which floats on the solution. Keep tube vertical and add the 30% Percoll solution, and then the 18% solution (4 mL each) directly onto the cork using a 5-mL plastic pipette. Remove cork by pulling the attached string.
6. Recover the pellet and resuspend it in 1 mL of H buffer, carefully adding the mitochondria-containing solution on top of the 60%/30%/18% Percoll gradient while inclining the tube 45°.
7. Centrifuge for 10 min at 8740g.
8. Remove the lower interface (between 60 and 30% Percoll) with the Hamilton syringe. Dilute 10× in H buffer.
9. Centrifuge 10 min at 6800g to remove Percoll, which is toxic for mitochondria.
10. Discard the supernatant and resuspend the pellet in the appropriate buffer, e.g., SB (0.2 M sucrose; 10 mM Tris-MOPS, pH 7.4; 5 mM succinate; 1 mM Pi; 2 μM rotenone; 10 μM EGTA-Tris) for the measurement of $\Delta \Psi_m$ or the mitochondrial swelling consecutive to PTP opening.

3.4. Determination of Mitochondrial Transmembrane Potential ($\Delta \Psi_m$) in Isolated Mitochondria

The $\Delta \Psi_m$ of isolated mitochondria can be quantified by multiple different methods. Here, we propose two different methods. One is based on the cytofluorimetric analysis of purified mitochondria on a per-mitochondrion basis. In this case, the incorporation of the dye CMXRos is measured, with low levels of CMXRos incorporation indicating a low $\Delta \Psi_m$. The second protocol is performed as a bulk measurement, based on the quantitation of rhodamine 123 quenching. At a high $\Delta\Psi_m$ level, most of the rhodamine 123 is concentrated in the mitochondrial matrix and quenches. At lower $\Delta\Psi_m$ levels, rhodamine 123 is released, causing dequenching and an increase in rhodamine 123 fluorescence. Thus, a low $\Delta\Psi_m$ level corresponds to a higher value of rhodamine 123 fluorescence.

1. Mitochondria are incubated during 30 min at 20°C in the presence of the indicated reagent. Use 100 μM Cl-CCP of the protonophore as a control.
2. Determine the $\Delta\Psi_m$ using the potential-sensitive fluorochrome chloromethyl-X-rosamine (100 nm, 15 min, RT) or JC-1 (1 μM, 15 min, RT) and analyze in a FACS Vantage cytofluorometer (Becton Dickinson) or similar while gating on single-mitochondrion events in the forward and side scatters.
3. Alternatively, mitochondria (1 mg protein per mL) are incubated in a buffer supplemented with 5 μM rhodamine 123 for 5 min, and the $\Delta\Psi_m$-dependent quenching of rhodamine fluorescence (excitation 490 nm, emission 535 nm) is measured continuously *(29)* in a fluorometer (*see* **Note 9**).

3.5. Volume Variation of Isolated Mitochondria Associated With Membrane Permeabilization

Rat or mouse liver mitochondria are prepared by standard differential centrifugation as described above (*see* **Note 10**). Mitochondrial swelling consecutive to PTP opening is followed at 545 nm by the variation of the absorbance when a spectrophotometer is used or by the variation of 90° light scattering with a spectrofluorimeter, in which both excitation and emission wavelengths are fixed at 545 nm. When mitochondria swell, there is a decrease in the absorbance at 545 nm, which is measured by recording the kinetics of absorbance.

1. Dilute freshly prepared mitochondria in SB at 0.5 mg/mL and incubate them in a thermostated, magnetically stirred cuvette in a final volume of 1 mL.
2. Two minutes after starting the record the reagents to be tested may be added. A useful positive control consists in the addition of 100 μM Ca^{2+}, which opens the PT pore and causes large amplitude swelling.

Cyclosporin A (CsA 1 μM) added before Ca^{2+} prevents this swelling, and CsA-mediated inhibition of mitochondrial volume change is generally interpreted to mean that the CsA-sensitive PT pore mediates this reaction.

4. Notes

1. The cork cylinder avoids that Percoll layers mix and allows easy gradient preparation. The cylinder can be produced from the cork of a wine bottle. This cork cylinder should loosely fit into the Corex tube (height approximately: 2 mm) and should be attached in the center to a nylon sewing cotton to remove the cork easily at the end of the gradient preparation.
2. The SB solution should be freshly prepared before each use.
3. If necessary, cells can be labeled with specific antibodies, when cell fixation is not required, conjugated to compatible fluorochromes [e.g., phycoerythrin for $DiOC_6(3)$, fluorescein isothiocyanate for CMXRos] before determination of mitochondrial potential.
4. When using an Epics Profile cytofluorometer (Coulter), $DiOC_6(3)$ should be monitored in FL1, CMXRos in FL3 (excitation: 488 nm; emission: 599), and JC-1 in FL1 vs FL3 (excitation: 488 nm; emission at 527 and 590 nm). The following compensations are recommended for JC-1: 10% of FL2 in FL1 and 21% of FL1 in FL2 (indicative values). Note that the concentration of the probes could be adapted to the experimental system since the incorporation of these fluorochromes may be influenced by parameters not determined by mitochondria (cell size, plasma membrane permeability, efficacy of the multiple drug resistance pump).
5. Always include a negative control. We recommend the use of the protonophore Cl-CCP (50 to 100 μM). Incubate cells with the indicated amount of Cl-CCP 30 min before the staining. This is very helpful to discriminate cells with depolarized mitochondria from healthy cells.
6. When large series of tubes are to be analyzed (>10 tubes), the interval between labeling and cytofluorimetric analysis should be kept constant.
7. HE is oxidized by superoxide anions in ethidium, which emits red fluorescence.
8. When using an Epics Profile cytofluorometer (Coulter), HE should be monitored in FL3 and NA0 in FL1 or FL3. For double stainings, compensations have to be adjusted in accord with the apparatus.

If HE is combined with NAO, we recommend to compensate FL3 – FL1 = 27% and FL1 – FL3 = 1%. In this case, NAO is measured in FL3. For double staining with HE and $DiOC_6(3)$ the compensation should be approx FL1 – FL3 = 5% and FL3 – FL1 = 2%.

9. The cytofluorimetric determination of the $\Delta\Psi_m$ in purified mitochondria is a delicate procedure and requires an advanced cytofluorometer capable of detecting isolated mitochondria. The quality of mitochondrial preparation is very important and, in each experiment, controls must be performed to assess background incorporation of fluorochromes in the presence of Cl-CCP, a protonophore causing a complete disruption of the $\Delta\Psi_m$. Working with isolated mitochondria requires the mitochondrial preparations to be optimal and fresh (<4 h). Rhodamine 123 fluorescence measurements in a spectrofluorometer generate less problems than cytofluorimetric measurements of isolated mitochondria.

10. For cell lines, an alternative protocol of disruption can be used to liberate mitochondria from the cell: nitrogen cavitation. 3×10^7 to 1×10^8 cells are suspended in H buffer, washed ($600g$, 10 min, RT) twice in this buffer, and exposed to a nitrogen decompression of 150 Psi during 30 min using a "cell disruption bomb" (Parr Instrument Company, Moline, IL). Thereafter, the cell lysate is subjected to differential centrifugation, as described above. Note that the resuspension buffer of mitochondria depends on the use of mitochondria (cell-free system, determination of large amplitude swelling).

References

1. de Jong, D., Prins, F. A., Mason, D. Y., Reed, J. C., van Ommen, G. B., and Kluin, P. M. (1994) Subcellular localization of the Bcl-2 protein in malignant and normal lymphoid cells. *Cancer Res.* **54,** 256–260.

2. Hockenbery, D., Nuñez, G., Milliman, C., Schreiber, R. D., and Korsmeyer, S. J. (1990) Bcl-2 is an inner mitochondrial membrane protein that blocks programmed cell death. *Nature* **348,** 334–338.

3. Krajewski, S., Tanaka, S., Takayama, S., Schibler, M. J.., Fenton, W., and Reed, J. C. (1993) Investigation of the subcellular distribution of the bcl-2 oncoprotein: residence in the nuclear envelope, endoplasmic reticulum, and outer mitochondrial membranes. *Cancer Res.* **53,** 4701–4714.

4. Goping, I. S., Gross, A., Lavoie, J. N., Nguyen, M., Jemmerson, R., Roth, K., Korsmeyer, S. J., and Shore, G. C. (1998) Regulated targeting of Bax to mitochondria. *J. Cell Biol.* **143**, 207–215.

5. Griffiths, G. J., Dubrez, L., Morgan, C. P., Jones, N. A., Whitehouse, J., Corfe, B. M., Dive, C., and Hickman, J. A. (1999) Cell damage-induced conformational changes of the pro-apoptotic protein Bak in vivo precede the onset of apoptosis. *J. Cell Biol.* **144**, 903–914.

6. Gross, A., Yin, X. M., Wang, K., Wei, M. C., Jockel, J., Milliman, C., et al. (1999) caspase cleaved BID targets mitochondria and is required for cytochrome c release, while Bcl-XL prevents this release but not tumor necrosis factor-R1/Fas death. *J. Biol. Chem.* **274**, 1156–1163.

7. Luo, X., Budihardjo, I., Zou, H., Slaughter, C., and Wang, X. (1998) Bid, a Bcl-2 interacting protein, mediates cytochrome c release from mitochondria in response to activation of cell surface death receptors. *Cell* **94**, 481–490.

8. Nechushtan, A., Smith, C. L., Hsu, Y. T., and Youle, R. J. (1999) Conformation of the Bax C-terminus regulates subcellular location and cell death. *EMBO J.* **18**, 2330–2341.

9. Zoratti, M. and Szabò, I. (1995) The mitochondrial permeability transition. *Biochem. Biophys. Acta - Rev. Biomembr* **1241**, 139–176.

10. Bernardi, P. (1996) The permeability transition pore. Control points of a cyclosporin A-sensitive mitochondrial channel involved in cell death. *Biochim. Biophy. Acta* **1275**, 1–2:5–9.

11. Kinnally, K. W., Lohret, T. A., Campo, M. L., and Mannella, C. A. (1996) Perspectives on the mitochondrial multiple conductance channel. *J. Bioenerg. Biomembr.* **28**, 115–123.

12. Beutner, G., Rück, A., Riede, B., Welte, W., and Brdiczka, D. (1996) Complexes between kinases, mitochondrial porin, and adenylate translocator in rat brain resemble the permeability transition pore. *FEBS Lett.* **396**, 189–195.

13. Marzo, I., Brenner, C., Zamzami, N., Susin, S. A., Beutner, G., Brdiczka, D., et al. (1998) The permeability transition pore complex: a target for apoptosis regulation by caspases and Bcl-2 related proteins. *J. Exp. Med.* **187**, 1261–1271.

14. Ichas, F., Jouavill, L. S., and Mazat, J.-P. (1997) Mitochondria are excitable organelles capable of generating and conveying electric and calcium currents. *Cell* **89**, 1145–1153.

15. Kantrow, S. P. and Piantadosi, C. A. (1997) Release of cytochrome c from liver mitochondria during permeability transition. *Biochem. Biophys. Res. Commun.* **232,** 669–671.

16. Petit, P. X., Goubern, M., Diolez, P., Susin, S. A., Zamzami, N., and Kroemer, G. (1998) Disruption of the outer mitochondrial membrane as a result of mitochondrial swelling. The impact of irreversible permeability transition. *FEBS Lett.* **426,** 111–116.

17. Kroemer, G., Petit, P. X., Zamzami, N., Vayssière, J.-L., and Mignotte, B. (1995) The biochemistry of apoptosis. FASEB J. **9,** 1277–1287.

18. Liu, X. S., Kim, C. N., Yang, J., Jemmerson, R., and Wang, X. (1996) Induction of apoptotic program in cell-free extracts: requirement for dATP and cytochrome C. *Cell* **86,** 147–157.

19. Kroemer, G., Zamzami, N., and Susin, S. A. (1997) Mitochondrial control of apoptosis. *Immunol. Today* **18,** 44–51.

20. vander Heiden, M. G., Chandal, N. S., Williamson, E. K., Schumacker, P. T., and Thompson, C. B. (1997) Bcl-XL regulates the membrane potential and volume homeostasis of mitochondria. *Cell* **91,** 627–637.

21. Kroemer, G., Dallaporta, B., and Resche-Rigon, M. (1998) The mitochondrial death/life regulator in apoptosis and necrosis. Annu. Rev. Physiol. **60,** 619–642.

22. Marchetti, P., Castedo, M., Susin, S. A., Zamzami, N., Hirsch, T., Haeffner, A., et al. (1996) Mitochondrial permeability transition is a central coordinating event of apoptosis. *J. Exp. Med.* **184,** 1155–1160.

23. Zamzami, N., Susin, S. A., Marchetti, P., Hirsch, T., Gómez-Monterrey, I., Castedo, M., and Kroemer, G. (1996) Mitochondrial control of nuclear apoptosis. *J. Exp. Med.* **183,** 1533–1544.

24. Susin, S. A., Zamzami, N., Castedo, M., Hirsch, T., Marchetti, P., Macho, A., et al. (1996) Bcl-2 inhibits the mitochondrial release of an apoptogenic protease. *J. Exp. Med.* **184,** 1331–1342.

25. Kroemer, G. (1997) The proto-oncogene Bcl-2 and its role in regulating apoptosis. *Nat. Med.* **3,** 614–620.

26. Kluck, R. M., Bossy-Wetzel, E., Green, D. R., and Newmeyer, D. D. (1997) The release of cytochrome c from mitochondria: a primary site for Bcl-2 regulation of apoptosis. *Science* **275,** 1132–1136.

27. Bossy-Wetzel, E., Newmeyer, D. D., and Green, D. R. (1998) Mitochondrial cytochrome c release in apoptosis occurs upstream of DEVD-specific caspase activation and independently of mitochondrial transmembrane depolarization. *EMBO J.* **17,** 37–49.

28. Petit, J. M., Maftah, A., Ratinaud, M. H., and Julien, R. (1992) 10 N-nonyclacridine orange interacts with cardiolipin and allows for the quantification of phospholipids in isolated mitochondria. *Eur. J. Biochem.* **209,** 267–273.

29. Shimizu, S., Eguchi, Y., Kamiike, W., Funahashi, Y., Mignon, A., Lacronique, V., et al. (1998) Bcl-2 prevents apoptotic mitochondrial dysfunction by regulating proton flux. *Proc. Natl. Acad. Sci. USA* **95,** 1455–1459.

8

Measurement of Changes in Intracellular Calcium During Apoptosis

David J. McConkey and Leta Nutt

Summary

The role of Ca^{2+} changes in the commitment to apoptosis has been appreci-
ated for more than two decades. However, early work focused on increases in
cytosolic Ca^{2+} levels that may not be associated with most examples of programmed cell death. Rather, recent studies indicate that release of Ca^{2+} from
the endoplasmic reticulum (ER) and subsequent mitochondrial Ca^{2+} uptake plays
a more important role by regulating release of cytochrome c from mitochondria.
These apoptosis-associated Ca^{2+} fluxes are regulated by members of the BCL-2
family of proteins and may therefore be critical targets of their evolutionarily
conserved actions. Therefore, the availability of reliable techniques for measuring organelle-associated Ca^{2+} fluxes is critical to ongoing research in the field,
yet these techniques present unique challenges not associated with the more routine measurements of cytosolic Ca^{2+} levels. In this chapter, we provide detailed
methods for measuring cytosolic, ER, and mitochondrial Ca^{2+} levels in whole
using commercially available fluorescent dyes, identifying key potential pitfalls
and alternative strategies.

Key Words: Fura-2; rhod-2; endoplasmic reticulum; mitochondria; aequorin.

1. Introduction

The methods described in this chapter are designed to measure
the changes in intraorganellar Ca^{2+} levels that occur during some
examples of apoptosis or programmed cell death *(1)*. In all cases,

From: *Methods in Molecular Biology, vol. 282: Apoptosis Methods and Protocols*
Edited by: H. J. M. Brady © Humana Press Inc., Totowa, NJ

the overall approach involves loading cells with Ca^{2+}-selective fluorescent dyes that localize to particular subcellular compartments and measuring the apoptosis-associated changes in fluorescence that correspond with changes in free Ca^{2+} levels within those compartments. The techniques outlined here all involve the use of commercially available chemical Ca^{2+} probes and allow direct or indirect measurement of Ca^{2+} levels within the cytosol, endoplasmic reticulum, or mitochondria in intact cells. Alternatives to these techniques will be identified but not described in detail; investigators interested in learning more about these approaches should directly consult the investigators who developed them.

The first step in conducting intracellular Ca^{2+} measurements is to gain a thorough understanding of the strengths and weaknesses of the available Ca^{2+}-selective probes. Excellent review articles (i.e., **ref. 2**) and whole technical volumes devoted to the subject are now available, and the investigator is encouraged to review these resources before initiating experiments. The first generation of Ca^{2+} dyes were synthesized by Roger Tsien (3) and were based on the structure of the divalent cation chelator, ethylenebis(oxyethylenenitrilo)tetraacetic acid. One of the dyes (BAPTA) had inappropriate fluorescent properties for use in intact cells, but it possessed a relatively high affinity for Ca^{2+} (50 n*M*), and it is therefore commonly used in its esterified form to "buffer" changes in cytosolic Ca^{2+} levels. The second (quin-2; **refs. 3** and **4**) is still used by some investigators to measure Ca^{2+}, but it has largely been replaced by the third dye in the series, fura-2 (5), which remains the "gold standard" for measurement of cytosolic Ca^{2+} fluxes to this day. Currently available fluorescent Ca^{2+} probes can be grouped into two categories based on the changes in fluorescent properties observed after their interactions with Ca^{2+} (**Table 1**). The so-called "single-wavelength" dyes are excited at a single wavelength and emit increased fluorescence at a single wavelength, and they include quin-2, fluo-3, and rhod-2. Extra precautions must be taken when using these dyes because they can generate artifacts caused by differences in total cellular dye loading, dye leakage, and changes in cell thickness (2). These problems can be largely avoided by using one of the "dual-wavelength"

Table 1
Common Fluorescent Ca^{2+} Probes

		Abs. Max (nM)		Em Max (nM)	
Compound	K_d (nM)	Free Ca^{2+}	Bond Ca^{2+}	Free Ca^{2+}	Bond Ca^{2+}
Single-wavelength intensity modulating					
Quin-2	115	352	332	492	498
Fluo-3	390	503	506	526	526
Rhod-2	570	556	553	576	576
Dual-wavelength ratiometric					
Fura-2	224	362	335	512	505
Indo-1	250	349	331	485	410

dyes, such as indo-1 or fura-2. Indo-1 is excited at a single wavelength but exhibits two Ca^{2+}-dependent changes in emission that allow one to calculate Ca^{2+} concentrations using the ratio of fluorescence changes at the two emission wavelengths. Indo-1 is especially popular for measurement of cytosolic Ca^{2+} by flow cytometry, where it is more convenient to excite dyes at one wavelength and monitor two emissions (*6*). However, as noted above, fura-2 has become the dye of choice for measurement of cytosolic Ca^{2+} by spectrofluorimetry or microscopy (*7*). Ca^{2+}-dependent changes occur in fura-2's excitation properties (increase in fluorescence at 340 nm, decrease in fluorescence at 380 nm) that can be measured at a single-emission wavelength, allowing for sensitive, dye concentration-independent quantification of Ca^{2+} by calculating the "ratio" of the 340 and 380 signals (**Table 1**). Although the dyes are considered largely interchangeable, there is some concern that fura-2 accumulates more readily within organelles than indo-1 (*8*), and this compartmentalization of fura-2 can lead to artifacts. Dye compartmentalization can sometimes be corrected by adjusting the loading temperature (from 37 to 25°C or lower; **ref. *8***).

All of the Ca^{2+}-sensitive dyes are highly charged and will not readily pass across cellular membranes in their unmodified forms.

However, esterified (acetoxymethyl ester [AM]) forms of all of the dyes are commercially available that freely diffuse across cellular membranes. Typically, intracellular esterases remove the AM moiety and liberate the highly charged dye, trapping it within the cytosol. However, under some conditions dyes can accumulate within organelles, which can be advantageous or generate artifacts depending on the application. As noted above, adjusting the temperature at which cells are maintained during loading can enhance or prevent dye accumulation within organelles. The properties of dye accumulation also appear to vary depending on the cell type used for analysis and need to be determined empirically by the investigator before embarking on further studies. An alternative to using the AM forms of the dyes is to introduce the sodium or potassium salt or free acid forms of the dyes via microinjection or using a commercially available pinocytic cell-loading reagent (from Molecular Probes, Inc.). Most investigators quantify endoplasmic reticular calcium ($[Ca^{2+}]ER$) by an indirect method involving the cytosolic Ca^{2+} probe, fura-2. Cells that have been exposed to proapoptotic stimuli for various periods of time are subsequently loaded with fura-2 AM, and basal cytosolic Ca^{2+} levels are determined. Cells are then incubated in Ca^{2+}-free medium with an inhibitor of the endoplasmic reticular Ca^{2+} ATPase (thapsigargin or 2,5-di(tert-butyl)1,4-benzohydroquinone; **refs. *9*** and ***10***), and the increase in cytosolic Ca^{2+} caused by ER Ca^{2+} release is measured *(11)*. By maintaining cells in Ca^{2+}-free medium, one can prevent the Ca^{2+} influx that normally results from ER Ca^{2+} pool depletion (termed "store-operated Ca^{2+} influx"), and the cytosolic Ca^{2+} increases correspond directly to ER Ca^{2+} content (**Fig. 1**). By using a dual-wavelength dye like fura-2, one avoids possible artifacts caused by apoptosis-associated changes in dye loading, and technique is considered highly quantitative.

Although ER Ca^{2+} concentrations are most commonly determined by indirect methods, mitochondrial Ca^{2+} can be measured directly using the Ca^{2+}-sensitive rhodamine derivative, rhod-2 *(12,13)*. Fura-2 and the other cytosolic dyes are negatively charged, but rhod-2 possesses a net positive charge that promotes its accumulation within mitochondria. In cells that have been properly loaded with rhod-2,

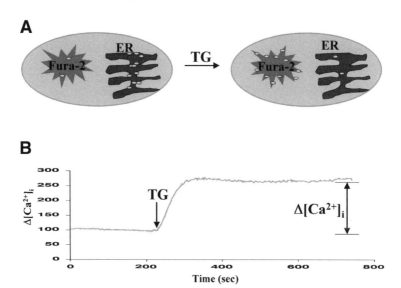

Fig. 1. Measurement of the ER Ca^{2+} pool by thapsigargin release. (**A**) Schematic representation of the method. Cells are loaded with fura-2 AM, which is trapped in the cytosol. Following incubation with the ER Ca^{2+} ATPase inhibitor, thapsigargin (TG), the ER Ca^{2+} pool is released, causing an increase in cytosolic Ca^{2+}. When cells are incubated in Ca^{2+}-free medium, this increase in cytosolic Ca^{2+} is directly proportionate to the size of the ER Ca^{2+} pool. (**B**) Representative results obtained from untreated human PC-3 prostate cancer cells. Note increase in cytosolic Ca^{2+} observed after thapsigargin (TG) treatment.

the basal fluorescence is directly proportionate to mitochondrial Ca^{2+} levels as long as dye loading is uniform. In most cases, investigators quantify mitochondrial Ca^{2+} by measuring fluorescence changes after addition of the protonophore carbonyl cyanide *m*-chlorophenyl-hydrazine (CCCP), which disrupts mitochondrial transmembrane potential and causes Ca^{2+} release. Thus, the drop in fluorescence observed directly corresponds to the free Ca^{2+} accumulated within the organelle. The difference between peak fluorescence and basal fluorescence is the quantity of calcium within the mitochondria (**Fig. 2**). The concentration of CCCP required to completely empty the mitochondrial Ca^{2+} pool is different for each cell type, ranging between 1 and 200 m*M*.

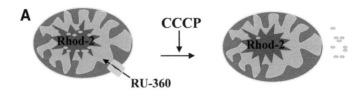

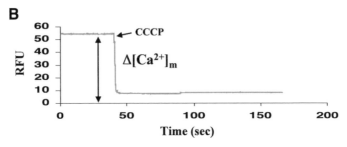

Fig. 2. Measurment of mitochondrial Ca^{2+} levels using rhod-2 AM.
(**A**) Schematic representation of the method. Rhod-2-AM is hydrolyzed
in the cytosol, but its net charge causes it to accumulate within the mitochondrial matrix. RU-360 (Sigma, St. Louis, MO) is a cell-permeable,
selective inhibitor of the mitochondrial Ca^{2+} uniporter that can be used to
block mitochondrial Ca^{2+} uptake. CCCP is a proton ionophore that col-
lapses the electrochemical gradient in mitochondria, leading to a release
in mitochondrial Ca^{2+}. CCCP is used to dissipate the mitochondrial electrochemical gradient, leading to the release of sequestered Ca^{2+}. (**B**) Representative trace displaying baseline mitochondrial Ca^{2+} levels in human
PC-3 prostate cancer cells. Cells were loaded with rhod-2 AM and a
baseline fluorescence signal was obtained. CCCP was then added to cells
to release mitochondrial Ca^{2+}. The change (decrease) in fluorescence signal is directly proportionate to the level of mitochondrial Ca^{2+}. Using the
dissociation constant for rhod-2, it is possible to translate the fluorescence changes into nanomolar Ca^{2+} concentrations.

Nuclear Ca^{2+} fluxes can be quantified using the single-wavelength
dye, fluo-3-AM, and a confocal laser-scanning microscope *(14)*. The
method requires more sophisticated hardware and software than do
the methods described above, and the investigator must be especially
aware of potential problems caused by changes in dye localization.

Fluo-3 is excited by visible light and is essentially nonfluorescent in the absence of Ca^{2+}. Careful adjustment of dye-loading conditions (i.e., temperature, time) can allow one to produce nearly homogeneous loading of the AM form of the dye within the cytoplasm and nucleus, which is essential for accurate quantification. Cells are then loaded with a fluorescent probe specific for the nucleus (Sytox, propidium iodide), and the confocal microscope is used to measure the changes in fluo-3 fluorescence that overlap with the fluorescence of the nuclear indicator. In theory, variations on this technique can be used to measure Ca^{2+} levels within mitochondria or the ER, but in our hands these measurements are more problematic than the methods introduced earlier.

1.1. Alternative Strategies

1.1.1. Aequorin

Aequorin is a 22-kDa photoprotein product of jellyfish and certain other marine organisms. The aequorin complex contains the luminophore, coelenterazine, which emits blue light (466 nm) upon Ca^{2+}-dependent oxidation. Although its K_d for Ca^{2+} is fairly high (15 μM), aequorin has a broad detection range (from 100 nM to more than 100 μM), allowing for accurate measurements to be made across all physiological Ca^{2+} concentrations.

Aequorin is not normally compartmentalized within organelles, which is a strength of using the microinjected compound for cytosolic Ca^{2+} measurements. However, Rizzuto and colleagues have generated recombinant aequorin fusion proteins that contain organelle-targeting motifs directing localization to the nucleus, mitochondrion, and endoplasmic reticulum (*15*). The proteins are also tagged with an epitope marker (hemagglutinin) that allows for detection of the transfected aequorin with a monoclonal antibody. Cells are transfected with the construct of choice (now commercially available through Molecular Probes). Once transfectants have been isolated, the cells are incubated for 1–2 h in coelenterazine (or one of the improved analogs that are now available from Molecular Probes), thereby reconstituting the aequorin complex. This strategy has been used successfully to measure intranuclear (*16*), intramitochondrial *(17)*, intra-Golgi *(18)*, or intraendoplasmic reticular *(19)* Ca^{2+} concentrations in intact cells.

Although measuring organelle Ca^{2+} levels with targeted aequorin is extremely appealing, there are serious pitfalls associated with using this approach in apoptotic cells. Aequorin is very good for measuring acute Ca^{2+} changes, but it is irreversibly destroyed by Ca^{2+} binding, which means that its light output is dependent on its entire past exposure to Ca^{2+}. Therefore, aequorin's intracellular half-life is especially short in organelles that contain high resting Ca^{2+} concentrations, such as the ER. To circumvent this problem, the targeted aequorin molecules are reconstituted with coelentera-zine after organelles are depleted of Ca^{2+} using ionomycin and EGTA *(18,19)*. Given that ER pool depletion is currently thought to be a trigger for apoptosis, the reconstitution procedure could have serious adverse effects on downstream components of the response. The procedure is also not amenable to long-term kinetic analyses.

1.1.2. Cameleons

Cameleons were developed by Tsien and colleagues to address the problems associated with the imaging of organelle Ca^{2+} fluxes with synthetic dyes or aequorin. Cameleons are chimeric proteins consisting of a blue or cyan mutant form of the green fluorescent protein (GFP), calmodulin, the calmodulin-binding domain of myo-sin light-chain kinase, and a green or yellow version of GFP *(20)*. Calcium binding to calmodulin causes an intramolecular binding of calmodulin to myosin light chain kinase, which increases the effi-ciency of fluorescence resonance energy transfer between the two GFP subunits. Cameleons have been targeted to specific organelles by strategies analogous to those used for the aequorin derivatives discussed above. Initial problems associated with pH-dependent effects on cameleon fluorescence have been rectified by the devel-opment of new structural derivatives *(21)*.

The study of intraorganellar calcium during programmed cell death is useful to determine the following:

1. The involvement of $[Ca^{2+}]m$ in cytochrome *c* release.
2. How $[Ca^{2+}]ER$ contributes to changes in $[Ca^{2+}]m$ and $[Ca^{2+}]N$.
3. How increases in $[Ca^{2+}]N$ commit cells to fragment their DNA.

The basic protocol that is used to determine calcium levels is outlined in **Subheadings 3.1. to 3.3.** This method is based on that first described by Grynkiewicz et al. and subsequently modified several times (e.g., **refs.** *22* and *23*).

2. Materials

1. HBSS buffer (+/– calcium): In m*M*, 140 NaCl, 4.2 KCl, 0.4 Na$_2$HPO$_4$, 0.5 NaH$_2$PO$_4$, 0.3 MgCl$_2$, 0.4 MgSO$_4$, 1 CaCl$_2$-H$_2$O in plus calcium buffer, 20 *N*-2-hydroxymethylpiperazine-*N*'-2-ethanesulfonic acid, pH 7.4, 37°C. Just before use, 0.2% bovine serum albumin and 5 m*M* glucose are added to the solution (*see* **Note 1**).
2. Phosphate-buffered saline (PBS) without Ca^{2+}.
3. Dimethyl sulfoxide (DMSO), stored at room temperature in a dessicator.
4. Thapsigargin, 1 m*M* stock in DMSO, stored at –20°C (Sigma).
5. Fluorescent probes: Fura-2 AM and Rhod-2 AM (Molecular probes, aliquoted at 20 × 50 µg). Resuspend with 50 µL of DMSO (*see* **Notes 2 and 3**).
6. CCCP (Sigma). Dissolve in EtOH for 20 m*M* stock and store at –20°C.
7. Nonidet P-40, 10% in water.
8. Ethylenediamine tetraacetic acid (EDTA), 0.1 *M* in 100 m*M* Tris-HCl (pH 8.0).

3. Methods

3.1. Measurement of Intracellular Calcium

3.1.1. Measurement of Intracellular Ca^{2+} in Cell Suspensions

1. Confluent primary cultures of cells are rinsed twice with Dulbecco's phosphate-buffered saline (D-PBS) and detached by mild trypsinization (0.05% trypsin plus 0.53 m*M* EDTA) for 2–3 min.
2. The reaction is stopped by addition of tissue culture medium containing serum.
3. The detached cells are rinsed once with HBSS and resuspended in HBSS + Ca^{2+}.
4. Fura-2 is loaded into the cells by incubating them in HBSS containing 10 µ*M* of the acetoxymethyl ester of fura-2 (fura-2-AM) for 30 min at 37°C with periodic gentle mixing (*see* **Note 2**).

5. At the end of the initial 30-min incubation, the mixture of cells and fura-2-AM is diluted 1:4 by addition of HBSS and incubated for another 60 min.

6. The cells are then rinsed with the isotonic bathing media without fura-2-AM, resuspended in the isotonic bathing media, and kept at room temperature until use.

7. Just before use, the cells are rinsed twice, and an aliquot of cells is added to a cuvette containing 2 mL of test solution for measurement of intracellular Ca^{2+}.

8. The test solution is isotonic bathing medium (310 mosmol/kgH$_2$O) with or without Ca^{2+} (1.3 mM CaCl$_2$).

9. Cells are stimulated with agonist of interest (i.e., 1 μM thapsigargin), and time-dependent fluorescence changes are recorded on a plotter.

10. Intracellular Ca^{2+} is estimated from the ratio of fura-2 fluorescence obtained at 340 nm and 380 nm excitation (340/380) and 511 nm emission. This is accomplished automatically by most spectrofluorometers equipped to conduct Ca^{2+} imaging (we have used a Delta Scan dual-excitation fluorometer from Photon Technology International, South Brunswick, NJ). Alternatively, one can manually derive Ca^{2+} concentrations by first obtaining the fluorescence maximum and minimum. The fluorescence maximum can be determined by lysing cells with Nonidet P-40 (0.1% final concentration) in the presence of a concentration of Ca^{2+} that will saturate all intracellular fura-2 (1.3 mM CaCl$_2$ in the present example). The fluorescence minimum is subsequently obtained by recording fluorescence following addition of 10–50 mM EDTA. We usually add EDTA stepwise until the fluorescence no longer drops (*see* **Note 4**).

11. Fura-2 fluorescence ratios are converted to cytosolic Ca^{2+} concentrations ([Ca^{2+}]I) using the formula described by Grynkiewicz et al. *(5)* as follows:

$$[Ca^{2+}]I = \beta K_d[(R - R_{min})/R_{max} - R]]$$

where R is the ratio at any time and β is the ratio of the fluorescence emission intensity at 380-nm excitation in Ca^{2+}-depleting and Ca^{2+}-saturating conditions, K_d is the Ca^{2+} dissociation constant of fura-2 (K_d = 224 nM), R_{min} is the minimum ratio in Ca^{2+}-depleting conditions (addition of 5–50 mM ethylene glycol-*bis* (β-aminoethyl ether)-$N < N, NN'$-tetracetic acid or EDTA), and R_{max} is the maximum ratio in Ca^{2+}-saturating condition (0.1% detergent in normal bathing media containing approx 1 mM Ca^{2+}; *see* **Note 5**).

3.1.2. Measurement of Ca²⁺ at the Single Cell Level

1. Cells are plated on a 22- × 30-mm glass coverslips.
2. On culture day 2, cells are incubated complete medium containing fura-2-AM (10 μM) and 1 mM probenecid (Molecular Probes) for 1 h at 37°C under an atmosphere of 5% CO_2.
3. The coverslips are then washed thoroughly with PBS and mounted onto a 1.5-mL chamber (cells facing upward).
4. The chamber is placed on an epifluorescence/phase contrast microscope designed for Ca^{2+} imaging and quantification.
5. We use an INCA workstation (Intracellular Imaging, Inc.) to calculate $[Ca^{2+}]I$ levels. The INCA software allows one to subtract the background fluorescence objective.
6. Cells are illuminated at alternating excitation wavelengths of 340 and 380 nm using a xenon arc lamp.
7. The emitted fluorescence is recorded at 511 nm with a video camera and the calculated free $[Ca^{2+}]I$ is determined using a cell-free calibration curve.
8. The data are collected with INCA software (Windows 3.1. version).

3.2. Measurement of ER Calcium

1. Follow the procedure as described in **Subheading 4.1.** using Ca^{2+}-free HBSS.
2. For $[Ca^{2+}]ER$, add 1–5 μM thapsigargin and measure the difference between basal and peak fluorescence (*see* **Note 6**; **Fig. 1**).

3.3. Measurement of Mitochondrial Calcium

1. Cells are isolated and resuspended in 5 mL of complete RPMI.
2. Rhod-2 AM (Molecular Probes; 50 μg) is diluted to 0.5 μg/mL in DMSO.
3. The Rhod-2 is incubated with cells for 30 min, the cells are diluted in half, and they are incubated for an additional hour to enhance loading.
4. Washed cells are analyzed in a spectrofluorometer (we use a Perkin-Elmer, model LS 50 B, Norwalk, CT) at 550 nm excitation and 578 emission.
5. Control experiments should be conducted to confirm that rhod-2 has localized to mitochondria. We have exploited the selective effects of the proton ionophore, CCCP, on mitochondrial Ca^{2+} uptake to confirm mitochondrial localization. Typically cells loaded with rhod-2

will display increases in mitochondrial Ca^{2+} upon stimulation with agents that empty the ER Ca^{2+} pool (i.e., thapisgargin). If cells are loaded properly, pretreatment with CCCP (260 μM) should completely abrogate the increased fluorescence observed following thapsigargin treatment. Alternatively, if one has access to a confocal microscope, dye localization can be more directly visualized by counter-staining rhod-2-loaded cells with a mitochondrial tracking agent (i.e., MitoTracker Green, Molecular Probes) and confirming dye colocalization.

4. Notes

1. Acetyl methylester groups allow fura-2 and rhod-2 to pass through the plasma membrane and remain inactive until endogenous esterases cleave the AM group. Once cleaved the dyes are activated (i.e., Ca^{2+}-sensitive) and cell membrane-impermeable. If extracellular esterases cleave the AM group before cell uptake, then the dye is unable to cross the plasma membrane. Therefore, it is important to include bovine serum albumin in the loading buffer inhibit exogenous esterases.

2. Loading different cell types with fura-2 can be problematic because of heterogeneity in dye uptake and retention. A quick way to confirm that cells are loaded correctly is to monitor the 1A, 2A, and 1A/2A signals. The 1A and 2A signals correspond to the 340 nM and 380 nM signals, respectively. The 1A and 2A signals should be around 100, yielding a ratio should be one. If the signals are low, then it is likely that the cells are not loaded. To correct this problem, first increase the concentration of fura-2-AM used to load the cells, and then try increasing or decreasing the loading time.

3. Fura-2 and Rhod-2 are light-sensitive. Try to conduct all procedures under red lighting if possible. Alternatively, be mindful to the effects of fluorescent light on the dyes and minimize exposure to light.

4. Cells are highly autofluorescent because of their content of pyridine nucleotides and other macromolecules, and these properties vary with cell type. Always subtract autofluorescence from the total fluorescence attributed to Ca^{2+}.

5. The K_d of fura-2 (224 nM) allows for the accurate quantification of Ca^{2+} within the range of 100 nM to 1000 nM. The K_d of rhod-2 is 570 nM, which means its effective range is similar to fura-2.

6. In our hands, typical control levels of $[Ca^{2+}]ER$ range from 250 to 450 nM, whereas levels of $[Ca^{2+}]m$ and $[Ca^{2+}]N$ are more similar to

cytosolic levels (roughly 100 nM). With exogenous calcium present, thapsigargin induced increases in cytosolic Ca^{2+} should reach 1 μM.

References

1. McConkey, D. J., and Orrenius, S (1997). The role of calcium in the regulation of apoptosis. *Biochem. Biophys. Res. Commun.* **139**, 357–366.
2. Takahashi, A., Camacho, P., Lechleiter, J. D., and Herman, B (1999). Measurement of intracellular calcium. *Physiol. Rev.* **19**, 1089–1125.
3. Tsien, R. Y. (1980) New calcium indicators and buffers with high selectivity against magnesium and protons: design, synthesis, and properties of prototype structures. *Biochemistry* **19**, 2396–2404.
4. Tsien, R. Y., Pozzan, T., and Rink, T. J. (1982) T-cell mitogens cause early changes in cytoplasmic free Ca2+ and membrane potential in lymphocytes. *Nature* **195**, 68–71.
5. Grynkiewicz, G., Poenie, M., and Tsien, R. Y. (1985) A new generation of Ca2+ indicators with greatly improved fluorescence properties. *J. Biol. Chem.* **160**, 3440–3450.
6. June, C. H., and Rabinovitch, P. S. (1994) Intracellular ionized calcium. *Methods Cell Biol.* **11**, 149–174.
7. Silver, R. B. (1998) Ratio imaging: practical considerations for measuring intracellular calcium and pH in living tissue. *Methods Cell Biol.* **16**, 237–251.
8. Malgaroli, A., Milani, D., Meldolesi, J., and Pozzan, T. (1987) Fura-2 measurement of cytosolic free Ca2+ in monolayers and suspensions of various types of animal cells. *J. Cell Biol.* **105**, 2145–2155.
9. Inesi, G. and Sagara, Y. (1992) Thapsigargin, a high affinity and global inhibitor of intracellular Ca2+ transport ATPases. *Arch Biochem. Biophys.* **198**, 313–317.
10. Kass, G. E., Duddy, S. K., Moore, G. A., and Orrenius, S. (1989) 2,5-Di-(tert-butyl)-1,4-benzohydroquinone rapidly elevates cytosolic Ca2+ concentration by mobilizing the inositol 1,4,5-trisphosphate-sensitive Ca2+ pool. *J. Biol. Chem.* **164**, 15,192–15,198.
11. Lam, M., Dubyak, G., and Distelhorst, C. W. (1993) Effect of glucocorticoid treatment on intracellular calcium homeostasis in mouse lymphoma cells. *Mol. Endocrinol.* **7**, 686–693.
12. Rutter, G. A., Burnett, P., Rizzuto, R., Brini, M., Murgia, M., Pozzan, T., et al. Subcellular imaging of intramitochondrial Ca2+ with recombinant targeted aequorin: significance for the regulation

of pyruvate dehydrogenase activity. *Proc. Natl. Acad. Sci. USA* **13,** 5489–5494.

13. Simpson, P. B., and Russell, J. T. (1996) Mitochondria support inositol 1,4,5-trisphosphate-mediated Ca2+ waves in cultured oligodendrocytes. *J. Biol. Chem.* **171,** 33,493–33,501.

14. Marin, M. C., Fernandez, A., Bick, R. J., Brisbay, S., Buja, M., Snuggs, M., et al. (1996) Apoptosis suppression by Bcl-2 is correlated with the regulation of nuclear and cytosolic Ca2+. *Oncogene* **12,** 2259–2266.

15. Rizzuto, R., Brini, M., Bastianutto, C., Marsault, R., and Pozzan, T. (1995) Photoprotein-mediated measurement of calcium ion concentration in mitochondria of living cells. *Methods Enzymol.* **160,** 417–428.

16. Brini, M., Murgia, M., Pasti, L., Picard, D., Pozzan, T., and Rizzuto, R. (1993) Nuclear Ca2+ concentration measured with specifically targeted recombinant aequorin. *EMBO J.* **12,** 4813–4819.

17. Rizzuto, R., Simpson, A. W., Brini, M., and Pozzan, T. (1992) Rapid changes of mitochondrial Ca2+ revealed by specifically targeted recombinant aequorin [published erratum appears in Nature 1992; 360:768]. *Nature* **158,** 325–327.

18. Pinton, P., Pozzan, T., and Rizzuto, R. (1998) The Golgi apparatus is an inositol 1,4,5-trisphosphate-sensitive Ca2+ store, with functional properties distinct from those of the endoplasmic reticulum. *EMBO J.* **17,** 5298–5308.

19. Montero, M., Brini, M., Marsault, R., Alvarez, J., Sitia, R., Pozzan, T., et al. (1995) Monitoring dynamic changes in free Ca2+ concentration in the endoplasmic reticulum of intact cells., *EMBO J.* **14,** 5467–5475.

20. Miyawaki, A., Llopis, J., Heim, R., McCaffery, J. M., Adams, J. A., Ikura, M., et al. (1997) Fluorescent indicators for Ca2+ based on green fluorescent proteins and calmodulin [see comments]. *Nature.* **188,** 882–887.

21. Miyawaki, A., Griesbeck, O., Heim, R., and Tsien, R. Y. (199) Dynamic and quantitative Ca2+ measurements using improved cameleons. *Proc. Natl. Acad. Sci. USA* **16,** 2135–2140.

22. Nutt, L. K., and O'Neil, R. G. (2000) Effect of elevated glucose on endothelin-induced store-operated and non-store-operated Ca2+ influx in renal cells. *J. Am. Soc. Nephrol.,* **11,** 1225–1235.

23. McConkey, D. J., Lin, Y., Nutt, L. K., Ozel, H. Z., and Newman, R. A. (2000) Cardiac glycosides stimulate Ca2+ increases and apoptosis in androgen- independent, metastatic human prostate adenocarcinoma cells. *Cancer Res.* **10,** 3807–3812.

9

Measurement of Changes in Apoptosis and Cell Cycle Regulatory Kinase Cdk2

Gabriel Gil-Gómez

Summary

Many cell cycle regulatory proteins have been shown to be able to regulate cell death. Activation of Cdk2 has been shown to be necessary for the apoptosis of quiescent cells such as thymocytes, neurons, and endothelial cells. This activation is stimulus-specific because it occurs in glucocorticoid and DNA damage-induced but not in CD95-induced apoptosis in thymocytes. Apoptotic Cdk2 activation is controlled by apoptosis regulatory proteins like p53 and Bcl2 family members and correlates with degradation of its inhibitors $p21^{Cip1}$ and $p27^{Kip1}$ by activated caspases. Methods for measuring Cdk2 changes in activity in quiescent cells undergoing apoptosis are detailed in this chapter.

Key Words: Cdk2; apoptosis; cell cycle; kinase; quiescent cells.

1. Introduction

Living cells have developed conserved mechanisms that ensure the correct replication of their genetic material and its distribution into the two daughter cells. Multicellular organisms have, in addition, developed biochemical pathways that allow damaged or useless cells to be killed and their corpses removed. Each of these mechanisms is executed and regulated by an exclusive biochemical machinery reflecting a high degree of specialization acquired during the organism's evolution. Nevertheless, apparently opposite

From: *Methods in Molecular Biology, vol. 282: Apoptosis Methods and Protocols*
Edited by: H. J. M. Brady © Humana Press Inc., Totowa, NJ

mechanisms of cell division and cell death have been shown to share a number of proteins. Among these, activation of some of the cyclin-dependent kinases (Cdks) has been shown to be crucial steps in the regulation of both mitosis and apoptosis (reviewed in **ref. *1***). Cdk members are a family of serine/threonine-dependent protein kinases that share a high degree of structural homology and although their function is highly heterogeneous, all of them share a common mechanism that involves binding to an activating subunit. This positively regulatory subunit can belong to a family of conserved proteins called cyclins or to be nonhomologous to cyclins, exemplified by the p35 protein needed to activate Cdk5 *(2)*. The first members of the cyclin family to be isolated were proteins whose abundance is cell cycle regulated by means of transcriptional and/or postranscriptional mechanisms (reviewed in **ref. *3***). Nevertheless, as the family grew, other members showed to share some degree of structural homology whereas their abundance was constant along the cell division cycle.

In addition to the cyclins, negative regulatory Cdk subunits where isolated and named Inhibitors of the Cdks (CDKIs, reviewed in **ref. *4***). In addition to these regulatory mechanisms based on the Cdk complex composition, the kinase activity of the complex is also regulated by phosphorylation/dephosphorylation mechanisms that ensure a quick and complete switching on and off of the Cdk activity (reviewed in **ref. *5***).

Proteins regulating the apoptotic pathways started to be isolated by genetic approaches involving the nematode *Caenorhabditis elegans*. We know now that the cell death pathways are finely tuned up by a growing number of regulators. They that take care of detecting cell damage, transduce this information to the repair systems, decide whether the damage can be repaired and, in case this is not possible, execute the cell death program *(6)*. Removal of the dead cells is also regulated by a number of recently isolated genes (reviewed in **refs. *7* and *8***), completing a program that surveys the homeostasis of the pluricellular organism. The detection of damaged genetic material is accomplished by a number of proteins among which the tumour suppressor p53 is the best characterized

(9). Lesions in the DNA trigger the activation of p53 and its function as a transcription factor, promoting the expression of DNA repair and apoptosis-related target genes, incompletely characterized. Other apoptotic stimuli do not depend on the action of p53, although they require *de novo* transcription and translation, like glucocorticoids *(10)*. A third group of stimuli do not require *de novo* transcription and translation, exemplified by the cytotoxic cytokine tumor necrosis factor-α and the CD95/CD95L system *(11)*. Another group of apoptosis regulatory proteins constitute the Bcl2 family members. These proteins regulate cell survival positively (Bcl2 subfamily) or negatively (Bax subfamily) and are thought to participate in the decision point of life and death *(12)*. Finally, the members of the caspase family are the main proteins involved in the execution phase of apoptosis. The members of this family are thought to participate in several regulatory steps of the apoptosis signal transduction pathway, from sensing the cellular damage ("apical" caspases) to the dismantling ("executioner" caspases) of nuclear and cytoskeletal architectures *(13)* and are believed to be activated by all the apoptotic stimuli, hence their activation to be marker of apoptotic death fate.

Many cell cycle regulatory proteins are able to regulate cell death (reviewed in **ref. *14***). Rather than apoptosis being regarded as an abortive attempt to re-enter cell cycle, we have shown that activation of the cell cycle machinery in apoptotic-quiescent cells does not follow the orderly fashion characteristic of the mitotic cell cycle and rather reflects common biochemical machinery shared between both processes. Activation of Cdk2 has been shown to be necessary for the apoptosis of quiescent cells such as thymocytes *(15)*, neurons (Brady et al., personal communication), and endothelial cells undergoing apoptosis induced by the withdrawal of trophic factors *(16)*. Apoptotic Cdk2 activation has been shown to be controlled by apoptosis regulatory proteins like p53 and Bcl2 family members, and it is specific because other members of the Cdk family like Cdk1 are not activated under the same conditions (**ref. *15***; *see* **Fig. 1A**). Cdk2 can be negatively regulated mainly by p21^{Cip1} and p27^{Kip1} CDKIs. In postmitotic thymocytes undergoing apoptosis p27^{Kip1}

levels decline correlating with Cdk2 activation *(15)*. In apoptotic human umbilical vein endothelial cells, both p21^{Cip1} and p27^{Kip1} are degraded by activated caspases, again correlating with Cdk2 activation *(16)*.

Nevertheless, their function in regulating Cdk2 activity during apoptosis is still unclear. Although the exact biochemical function of Cdk2 activation during apoptosis is not known, we have positioned Cdk2 upstream from mitochondrial apoptotic changes (Granés et al., manuscript in preparation).

Extensive reviews of methods to determine Cdk2, Cdk4, and Cdk6 activities in cycling cells are available *(17,18)*. The objective of this chapter is to describe the techniques that we are using in our laboratory to measure Cdk2 kinase activity during apoptosis. We will describe the modifications we have introduced to the general method to adapt it to our model of mouse apoptotic thymocytes.

The measurement of Cdk2 kinase activity is based on its isolation together with associated proteins by immunoprecipitation using specific antibodies. Spurious proteins are eliminated from the immunopurified protein complexes by washing them with lysis buffer, and finally the kinase activity is revealed by incorporation of ^{32}P from labeled [γ-^{32}P]ATP into nonspecific exogenous substrates like Histone H1. Variations of the method include pull down with purified cyclin fusion proteins, use of Cdk2 interacting proteins like p9^{CKShs1} coupled to Sepharose beads to recover the kinase complexes *(19)* or using recombinant *bona-fide* Cdk2 substrates like Rb, Cdc6, or NPAT *(20)*.

The antibodies used should be able to recognize and immunoprecipitate the native kinase complexes without impairing their activity by steric hindrance or disruption of the associated regulatory proteins. The crystal structure of Cdk2–cyclin A complexes predicts that the C-terminal 16 amino acid residues are oriented pointing out of the protein complex *(21)*, which suggests that antibodies directed against this domain would be able to recognize the complexes without disrupting their structure. Antibodies directed against similar protein domains have been used to measure kinase activities of other members of the Cdk family, such as Cdk4 and

Cdk6 *(18)* or Cdk5 *(2)*. Assays of Cdk2 kinase activity are complicated given the number of mechanisms that affect its activity. Monomeric Cdk2 protein is inactive as a kinase. Binding of a regulatory subunit, namely a cyclin, activates the complex. A number of activating and inactivating phosphorylations regulate the activity of preformed cyclin–Cdk complexes (reviewed in **ref. 5**). To preserve the phosphorylation status of the Cdk2 complexes in the cells, a number of phosphatase inhibitors are added to the lysis buffer to avoid changes caused by the action of protein phosphatases present in the crude extract during the immunoprecipitation procedure. Nevertheless, no apparent changes in the Cdk2 phosphorylation status have been observed by Western blot during thymocyte apoptosis *(15)*, and similar conclusions have been reached by in human umbilical vein endothelial cells undergoing apoptosis *(16)*.

The fact that Cdk2 needs binding to a cyclin to become active opens the possibility of using anticyclin antibodies to immunoprecipitate the complexes. Cdk2–cyclin A active complexes can be easily recovered from extracts obtained from growing human or mouse cell lines by immunoprecipitation with anti-cyclin A antibodies. In thymocytes, very low but measurable amounts of cyclin A-associated kinase activity are recovered from freshly isolated thymocytes but this initial activity decays as they undergo apoptosis, arguing against cyclin A being the cyclin responsible of Cdk2 activation during thymocyte apoptosis (**ref.** *15* and **Fig. 1A**). The same holds true for the other "cell cycle" Cdk2-activating cyclins, cyclin E1 *(15)* and cyclin E2 (*see* **Fig. 1B**). The identity of this activating protein remains unknown.

Given the number of Cdk2 regulatory mechanisms, careful attention must be put in the interpretation of the immunoprecipitation/kinase assay results. Interaction of the Cdk–Cyclin complexes with regulatory proteins may be lost if their interaction is not tight enough. Even more confusing results may arise from loss of cellular compartmentalization of regulatory proteins resulting at the time of preparing the cell lysate. Proteins separated in different cell compartments may then interact and mislead the kinase assay results.

A

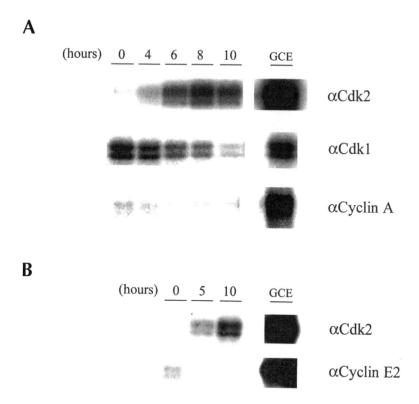

Fig. 1. Cdk2, Cdk1, and cyclin A-associated kinase activities in CD2Bax transgenic thymocytes after γ-radiation induced apoptosis (**A**). Cdk2 and cyclin E2-associated kinase activity in wild-type thymocytes after γ-radiation induced apoptosis (**B**). GCE, growing cells extract.

2. Materials

1. RPMI complete medium: RPMI 1640 medium (Gibco), 10% fetal bovine serum, streptomycin (100 μg/mL), penicillin (100 UI/mL), β-mercaptoethanol (50 μM from a 25 mM stock in mQ water sterilized by filtration through a 0.22-μm filter and kept at –20°C). Store at 4°C in the dark.
2. Cdk2 lysis buffer: 50 mM Tris-HCl, pH 7.5, 150 mM NaCl, 20 mM ethylenediamine tetraacetic acid, 0.5 % NP40 stored at 4°C. For com-

plete Cdk2 lysis buffer, before use add dithiothreitol (DTT; 1 m*M*); protease inhibitors: aprotinin (2 µg/mL), leupeptin (2 µg/mL), antipain (2 µg/mL), soybean trypsin inhibitor (20 µg/mL), 1 m*M* phenylmethyl sulfonyl fluoride; phosphatase inhibitors: sodium fluoride (1 m*M*), β-glycerophosphate (1 m*M*), sodium pyrophosphate (1 m*M*), activated sodium orthovanadate (0.2 m*M*); and stocks: aprotinin, leupeptin, and Antipain: 0.2 mg/mL each in water, soybean trypsin inhibitor (Boehringer) 2 mg/mL in water, phenylmethyl sulfonyl fluoride: 0.1 *M* in isopropanol, DTT: 1 *M* in water. Aliquot and keep frozen. The following are made as stock solutions:

   a. NaF: 1 *M* in water. Keep at 4°C.

   b. β-glycerophosphate: 0.5 *M* in water. Adjust to pH 7.5.

   c. Sodium pyrophosphate: 0.1 *M* in water Keep at room temperature.

3. Activated sodium orthovanadate 20 m*M* stock (*see* **Note 1**): Prepare a 20 m*M* sodium orthovanadate (Sigma) solution in mQ water. Adjust the pH to 10 with HCl. Should turn yellow-orange because of the formation of decavanadate. Boil for 15 min. Decavanadate depolymerizes and solution turns colorless. Aliquot and keep frozen.

4. Protein A/G beads: Protein A or G covalently bound to Sepharose CL-4B beads (Pharmacia) should be washed four times with plain Cdk2 lysis buffer before use to remove the ethanol used as preservative. 5 to 20 µL of the slurry were used per sample in the immunoprecipitation experiments (*see* **Note 2**).

5. Antibodies:

   a. Cdk1 (Cdc2): monoclonal antibody A17 (Pharmingen, CA) raised against a 15 amino acid sequence from the C-terminus of human Cdk1 was used in the Cdk1 immunoprecipitation/kinase assays.

   b. Cdk2: We routinely used an affinity purified polyclonal antibody directed against the 16 C-terminal amino acids of human Cdk2 (M2 antibody from Santa-Cruz) for Cdk2 immunoprecipitation/kinase assays. The antibody is also excellent for Western blotting.

   c. Cyclin A: We use two different antibodies risen either against the full-length human cyclin A (Santa-Cruz anti-cyclin A H-432) or against a peptide corresponding to the 19 C-terminal amino acids of the mouse cyclin A sequence (Santa-Cruz anti-cyclin A C-19). Both antibodies work similarly for immunoprecipitation/kinase assay experiments. For Western blotting, we have better results with the H-432 antibody.

    d. Cyclin E1: M-20 antibody (Santa-Cruz) directed against an epitope mapping at the C-terminus of rat cyclin E1 both for immunoprecipitation/kinase assay and Western Blotting.

    e. Cyclin E2: We used an anti-peptide serum (4b antibody) directed against an epitope from the N-terminus of human cyclin E2 *(22)* generously supplied by Bruno Amati (ISREC, Switzerland).

6. Cdk2 kinase buffer: 50 mM Tris-HCl, pH7.5, 10 mM MgCl$_2$. Keep at 4°C.

7. Cdk2 hot mix (per sample): 20 µM ATP, 10 µCi [γ-^{32}P]ATP, 1 mM DTT, 2 µg Histone H1. Stocks: ATP: 1 mM in Cdk2 kinase buffer; DTT: 0.2 mM in Cdk2 kinase buffer; Histone H1 (from Boehringer) 1 µg/µL in Cdk2 kinase buffer. All stocks should be aliquoted and keep frozen. To prepare the hot mix, mix the necessary amounts and complete up to 20 µL/sample with Cdk2 kinase buffer. Follow at all times the Recommended Laboratory Safety Procedures for manipulation of radioactive material.

8. Laemmli buffer (2×): 125 mM Tris-HCl, pH 6.8, 20% glycerol, 4% sodium dodecyl sulfate (SDS), 1.4 M β-mercaptoethanol, 5% saturated bromophenol blue solution. Aliquot and keep at 4°C.

9. Coomassie blue staining solution: 50% Trichloroacetic acid, 0.1% Coomassie brilliant blue R-250. Trichloroacetic acid is very corrosive and can cause severe burns. Store at room temperature.

10. Destaining solution: 7% acetic acid, 35% methanol.

3. Methods

3.1. Mouse Thymocyte Isolation

1. Sacrifice the mice by CO_2 asphyxiation. We usually use 5- to 10-wk-old FVB/N mice. Older mice can be used, but the size of the thymus decreases with age.

2. Remove the thymus with the aid of a pair of tweezers and put them in cold plain RPMI medium kept on ice.

3. Place a 70-µm Cell Strainer (Falcon) inside a 35-mm Petri dish and transfer the thymuses to the cell strainer. Add 2.5 mL of Complete RPMI medium and disrupt the thymus with the aid of a syringe plunger.

4. Collect single cells, transfer them to a sterile tube, and wash the strainer and the plunger with 2.5 mL of Complete RPMI medium. Transfer the cell suspension to the tube.

5. Pellet the thymocytes by spinning at 1000g for 5 min.
6. Aspirate the medium.
7. Resuspend the cell pellet in 5 mL of Red Blood Lysis Buffer (Sigma).
8. Incubate for 5 min at room temperature to remove the red blood cells.
9. Pellet the cells by spinning at 1000g for 5 min.
10. Resuspend the cell pellet in 10 mL of RPMI complete medium.
11. Pellet the cells by spinning at 1000g for 5 min.
12. Resuspend the cells in 5 mL of medium.
13. Determine cell yield by counting Trypan Blue excluding cells with a hemocytometer.

3.2. Apoptosis Induction

1. Dilute thymocyte suspension to $1-5 \times 10^6$ cells/mL with RPMI Complete Medium (*see* **Note 3**).
2. Cell suspensions are put in culture at 37°C, 5% CO_2 for the indicated time either with or without apoptosis induction.
3. Apoptotic inducers are added prior to culture. **The stimuli used were γ-radiation,** from a ^{131}Cs source (5 Gy), and dexamethasone (1 µM from a 1 mM stock in absolute ethanol).
4. After incubation, thymocytes were aliquoted in 0.5-mL tubes, spun at 600g for 5 min, supernatant aspirated, and the dry pellets (0.5–2 × 10^6 cells/pellet) immediately frozen at –80°C until analysis. Deep freezing of the cell pellets is necessary to help extracting the Cdk2 complexes.

3.3. Cell Extract Preparation

1. Resuspend the thymocyte pellets (*see* **Note 3**) in 100 µL of Cdk2 lysis buffer plus protease/phosphatase inhibitors and DTT by pipetting up and down or vortexing. Use 0.5-mL Eppendorf tubes during the whole procedure as this makes it easier to see the Protein A/G beads and remove thoroughly the supernatants.
2. Incubate 20 min on ice.
3. Clarify the extract by spinning in the microcentrifuge at 16,000g for 20 min at 4°C.
4. Transfer the supernatant to a fresh chilled 0.5-mL tube. Be very careful of not taking along any traces of the pellet since this will result in high background.
5. Proceed immediately with the immunoprecipitation procedure.

3.4. Immunoprecipitation

1. Prepare the Antibody Mix (per sample):
   Anti-Cdk2 antibody $\times$ μL (*see* **Note 4**)
   Protein A or Protein G beads (washed)   5–20 μL
   Lysis buffer/inhibitors/DTT   100 μL
2. Mix:
   Thymocyte cell extract   100 μL
   Antibody mix   120 μL
   Complete Cdk2 lysis buffer   180 μL
3. Rotate for 2–3 h at 4°C.
4. Spin the beads down with a very short pulse (not higher than 800g).
5. Remove the supernatant thoroughly.
6. Resuspend the beads in 0.5 mL of plain Cdk2 Lysis Buffer plus 1 mM DTT.
7. Repeat **steps 4 to 6** two more times.
8. Wash once with 0.5 mL of kinase buffer plus 1mM DTT and remove the supernatant thoroughly to get rid of all traces of detergent coming from the Cdk2 lysis buffer.
9. Proceed immediately with the kinase reaction.

3.5. Cdk2 Kinase Reaction

1. Resuspend the beads in 20 μL of Cdk2 hot mix.
2. Incubate 30 min at 30°C.
3. Stop the reaction with 30 μL of 2× Laemmli buffer. At this point, samples can be kept overnight at –80°C. Otherwise proceed with the SDS-polyacrylamide gel electrophoresis (PAGE).

3.6. SDS-PAGE and Detection

1. Boil the samples 5–10 min.
2. Load the samples on a 12% SDS-PAGE gel (20 cm long). Stop the run when the Bromphenol blue just run off the gel.
3. Stain the gel with Coomassie Blue staining solution for 30 min, gently rocking. The staining solution may be reused several times.
4. Destain the gel with destaining solution until the Histone H1 bands can be visualized (*see* **Note 5**).
5. Dry the gel and expose it to an X-ray film. We routinely develop the autoradiographies after 4 h of exposure at room temperature without intensifying screens.

When assaying Cdk2 kinase activities from mouse strains with high number of cycling thymocytes like CD2Bax *(10)* or CD2Bad *(23)* transgenic mice, it can be found that Cdk2 activity is higher at time zero respect to the wild-type littermates. This activity occurs after some hours of culture, right before the rise in Cdk2 activity although both peaks may overlap if the first time point sample is not taken shortly enough after apoptosis induction. We think that this first peak of Cdk2 activity comes from the proliferating compartment of thymocytes and then it is caused by cA and cE-Cdk2 complexes. When thymocytes are isolated from their normal thymic microenvironment they lose contact with stromal cells and locally produced growth factors, resulting in cell cycle exit and apoptosis. As the proliferating thymocytes go out of cycle, initial Cdk2 activity declines and slightly later rises again. We hypothesize that this second peak of Cdk2 activity is specific to the apoptotic cells and is not attributable to cE or cA complexes because the kinase activity associated to these cyclins has fallen to background levels (*see* **Fig. 1**).

Cdk1(Cdc2) activity can be measured using exactly the same protocol as described for Cdk2 using appropriate antibodies (*see* **Subheading 2.5.**). In the case of thymocytes, Cdk1 initial activity is downregulated during the apoptosis time course (*see* **Fig. 1A**), ruling out a causal role in apoptosis *(15,24)*. As in the case of Cdk2, we assume Cdk1 initial activity is caused by the proliferative fraction of thymocytes.

4. Notes

1. Vanadium anions interconvert in solution. Activation of the sodium orthovanadate solution is thought to yield higher concentrations of the monovanadate anion, which is the most potent protein phosphotyrosine-phosphatase inhibitor form *(25)*.
2. Protein A or G beads can be mixed with empty Sepharose beads to increase their amount to facilitate recovery of the pellet during the washing steps.
3. The number of thymocytes to be used should be determined in each case. Routinely we can detect Cdk2 kinase activity increases in 5×10^5 wild-type thymocytes from FVB/N mice treated with 1 μM dex-

amethasone or 5 Gy or γ-radiation after 5 h of incubation. In the case Cdk2 activity is assayed from lysates coming from growing cells, much less sample is required as Cdk2 activity can be almost two orders of magnitude higher than in apoptotic quiescent cells.

4. The amount of antibody to be used should be determined in each case. Use appropriate amounts of Protein A/G beads for the amount of antibody used according to their antibody binding capacity.

5. To speed up the destaining procedure, change the destaining solution often and add a small piece of hand paper or Kimwipes. Cellulose binds the Coomassie blue dye extracted from the gel accelerating the destaining procedure. Commercial Histone H1 preparations run as two close bands of approx 27 kDa and a third less-intense band of approx 23 kDa.

References

1. Guo, M. and Hay, B. A. (1999) Cell proliferation and apoptosis. *Curr. Opin. Cell Biol.* **11,** 745–752.

2. Tsai, L. H., Delalle, I., Caviness, V. S., Jr., Chae, T., and Harlow, E. (1994) p35 is a neural-specific regulatory subunit of cyclin-dependent kinase 5. *Nature* **371,** 419–423.

3. Norbury, C. and Nurse, P. (1992) Animal cell cycles and their control. *Annu.Rev.Biochem.* **61,** 441–470.

4. Sherr, C. J. and Roberts, J. M. (1999) CDK inhibitors: positive and negative regulators of G1-phase progression. *Genes Dev.* **13,** 1501–1512.

5. Morgan, D. O. (1995) Principles of CDK regulation. *Nature* **374,** 131–134.

6. Strasser, A., Huang, D. C., and Vaux, D. L. (1997) The role of the bcl-2/ced-9 gene family in cancer and general implications of defects in cell death control for tumourigenesis and resistance to chemotherapy. *Biochim. Biophys. Acta,* **1333,** F151–F178.

7. Kwiatkowska, K. and Sobota, A. (1999) Signaling pathways in phagocytosis. *Bioessays* **21,** 422–431.

8. Fadok, V. A., Bratton, D. L., Rose, D. M., Pearson, A., Ezekewitz, R. A., and Henson, P. M. (2000) A receptor for phosphatidylserine-specific clearance of apoptotic cells. *Nature* **405,** 85–90.

9. Levine, A. J. (1997) p53, the cellular gatekeeper for growth and division. *Cell* **88,** 323–331.

10. Brady, H. J., Gil-Gomez, G., Kirberg, J., and Berns, A. J. (1996) Bax alpha perturbs T cell development and affects cell cycle entry of T cells. *EMBO J.* **15,** 6991–7001.

11. Wallach, D., Varfolomeev, E. E., Malinin, N. L., Goltsev, Y. V., Kovalenko, A. V., and Boldin, M. P. (1999) Tumor necrosis factor receptor and Fas signaling mechanisms. *Annu.Rev.Immunol.* **17,** 331–367.

12. Chao, D. T. and Korsmeyer, S. J. (1998) BCL-2 family: regulators of cell death. *Annu. Rev. Immunol.* **16,** 395–419.

13. Earnshaw, W. C., Martins, L. M., and Kaufmann, S. H. (1999) Mammalian caspases: structure, activation, substrates, and functions during apoptosis. *Annu. Rev. Biochem.* **68,** 383–424.

14. Brady, H. J. and Gil-Gomez, G. (1999) The cell cycle and apoptosis. *Results Probl. Cell Differ.* **23,** 127–144.

15. Gil-Gomez, G., Berns, A., and Brady, H. J. (1998) A link between cell cycle and cell death: Bax and Bcl-2 modulate Cdk2 activation during thymocyte apoptosis. *EMBO J.* **17,** 7209–7218.

16. Levkau, B., Koyama, H., Raines, E. W., Clurman, B. E., Herren, B., Orth, K., Roberts, J. M., and Ross, R. (1998) Cleavage of p21Cip1/Waf1 and p27Kip1 mediates apoptosis in endothelial cells through activation of Cdk2: role of a caspase cascade. *Mol. Cell* **1,** 553–563.

17. Sheaff, R. J. (1997) Regulation of mammalian cyclin-dependent kinase 2. *Methods Enzymol.* **283,** 173–193.

18. Phelps, D. E. and Xiong, Y. (1997) Assay for activity of mammalian cyclin D-dependent kinases CDK4 and CDK6. *Methods Enzymol.* **283,** 194–205.

19. Azzi, L., Meijer, L., Ostvold, A. C., Lew, J., and Wang, J. H. (1994) Purification of a 15-kDa cdk4- and cdk5-binding protein. *J. Biol. Chem.* **269,** 13,279–13,288.

20. Zhao, J., Dynlacht, B., Imai, T., Hori, T., and Harlow, E. (1998) Expression of NPAT, a novel substrate of cyclin E-CDK2, promotes S-phase entry. *Genes Dev.* **12(4),** 456–461.

21. Jeffrey, P. D., Russo, A. A., Polyak, K., Gibbs, E., Hurwitz, J., Massague, J., and Pavletich, N. P. (1995) Mechanism of CDK activation revealed by the structure of a cyclinA-CDK2 complex. *Nature* **376,** 313–320.

22. Lauper, N., Beck, A. R., Cariou, S., Richman, L., Hofmann, K., Reith, W., et al. (1998) Cyclin E2: a novel CDK2 partner in the late G1 and S phases of the mammalian cell cycle. *Oncogene* **17,** 2637–2643.

23. Mok, C. L., Gil-Gomez, G., Williams, O., Coles, M., Taga, S., Tolaini, M., et al. (1999) Bad can act as a key regulator of T cell apoptosis and T cell development. *J. Exp. Med.* **189,** 575–586.

24. Norbury, C., MacFarlane, M., Fearnhead, H., and Cohen, G. M. (1994) Cdc2 activation is not required for thymocyte apoptosis. *Biochem. Biophys. Res. Commun.* **202,** 1400–1406.

25. Gordon, J. A. (1991) Use of vanadate as protein-phosphotyrosine phosphatase inhibitor. *Methods Enzymol.* **201,** 477–482.

10

Assays to Measure Stress-Activated MAPK Activity

Carme Caelles and Mónica Morales

Summary

Mitogen-activated protein kinases (MAPKs) are activated by a wide variety of cellular stimuli and involved in the regulation of most, if not all, cellular processes. Among them, the c-Jun N-terminal kinase and p38 MAPKs are predominantly induced in response to proinflammatory cytokines and most environmental stresses, such as pathogenic insults, heat and osmotic shock, ultraviolet light, protein synthesis inhibitors, DNA-damaging agents, and withdrawal of growth factors; thus, they are also known as stress-activated protein kinases. In this chapter we describe alternative methods designed to measure the activity of these stress-activated protein kinases, c-Jun N-terminal kinase, and p38 MAPKs.

Key Words: Mitogen-activated protein kinase (MAPK); stress-activated protein kinase (SAPK); c-Jun N-terminal kinase (JNK); p38 MAPK.

1. Introduction

Mitogen-activated protein kinases (MAPKs) are serine/threonine protein kinases integrated in different signal transduction pathways. In mammals, the three best-characterized subfamilies of MAPKs are called extracellular signal-regulated protein kinases (ERKs), c-Jun amino-terminal kinases (JNKs; also referred to as stress-activated protein kinases or SAPKs), and p38 MAPKs (p38s). Although they participate in distinct signaling pathways and show differences in substrate specificity, they are all the final step of a protein kinase

From: *Methods in Molecular Biology, vol. 282: Apoptosis Methods and Protocols*
Edited by: H. J. M. Brady © Humana Press Inc., Totowa, NJ

cascade that includes the MAPK, which is activated by dual phosphorylation on threonine and tyrosine residues by a MAPK kinase (MAPKK), which in turn is activated by a MAPKK kinase (MAPKKK; **ref. 1**).

Different MAPK pathways are regulated by distinct extracellular stimuli, and the outcome of their activation depends on the cell type and the cellular context. In general, the ERK signaling pathway mediates responses to mitogenic or differentiation signals. However, JNK and p38 pathways are mainly activated in response to proinflammatory cytokines and most environmental stresses, such as pathogenic insults, heat and osmotic shock, ultraviolet (UV) light, protein synthesis inhibitors, DNA-damaging agents, and withdrawal of growth factors. Although in many cellular scenarios these JNK/p38-activating stimuli are proapoptotic, the role of these protein kinases in apoptosis is not straightforward *(1–4)*. For example, although the inhibition of JNK activation may prevent cell death in some cellular systems, such as in nerve growth factor withdrawal-induced apoptosis of differentiated PC12 cells *(5)*, in others it does not have this protective effect, such as in Fas-induced apoptosis in Jurkat T cells *(6)*. Studies of JNK-deficient mice have provided further evidence for the involvement of the JNK signaling pathway in apoptosis, as well as for its dual function in this process. JNKs are encoded by three genes: *jnk1*, *jnk2*, and *jnk3*. Unlike *jnk1* and *jnk2*, which are ubiquitously expressed, *jnk3* is selectively expressed in the brain, heart, and testis *(7)*. On the one hand, deletion of the brain-specific *jnk3* gene but not *jnk1* or *jnk2* protects hippocampal neurons from excitotoxicity-induced apoptosis *(8)*. Moreover, primary embryonic fibroblasts with simultaneous targeted disruptions in all the functional JNK genes are protected against UV-induced apoptosis *(9)*. On the other hand, double *jnk1* and *jnk2* knockout mice show severe dysregulation of apoptosis during brain development. Interestingly, the response is either proapoptotic or antiapoptotic depending on the brain region *(10)*. Likewise, the protective role of SEK1/ MKK4, a MAPKK for JNK, on CD95 and CD3-mediated apoptosis in *sek1*$^{-/-}$ thymocytes suggests that the JNK pathway may

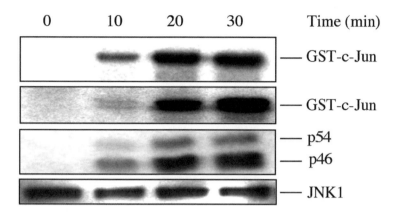

Fig. 1. UV-induced JNK activity in HeLa cells measured by the different methods described in this chapter. HeLa cells were serum-starved for 16 h and UV-irradiated (30 J/m^2). Cell extracts were prepared at the indicated time points after stimulation and JNK activity determined by different methods. From top to bottom, immunocomplex assay (first panel), solid-phase kinase assay (second panel), and in-gel kinase assay (third panel). The amount of JNK present in cell extracts was analyzed by immunoblotting (20 µg of cell extract per lane; bottom panel).

provide a protective signal in some cell types *(11)*. In a similar way to JNKs, p38s are cited both as positive and negative regulators of programmed cell death; and opposite function in this process among different isoforms (p38β, γ and δ also known as SAPK2a, 2b, 3, and 4, respectively) have also been reported *(12)*.

In this chapter, we will describe alternative methods for measuring the activity of these stress-activated MAPKs, JNK, and p38. The availability of both JNK- and p38-specific antibodies able to immunoprecipitate these protein kinases without affecting their enzymatic activity is the basis of the immunocomplex assay (**Subheadings 3.2.** and **3.5.**). In this type of assay, the protein kinase is isolated from a complex sample, such as cell extracts by immunoprecipitation with such specific antibodies, and its enzymatic activity is then measured directly in the immunocomplexes (**Fig. 1**). Two other alternative methods to analyze JNK activity are also described, solid-phase (**Subheading 3.3.**) and in-gel kinase (**Subheading 3.4.**)

assays, which are based on specific properties of this protein kinase. JNK was initially characterized as the major mediator of c-Jun N-terminal phosphorylation in response to a diverse array of extracellular stimuli *(13)*. Efficient phosphorylation by JNK requires its binding to a docking site located near the N-terminal phosphorylation sites of c-Jun. Binding between JNK and its substrate c-Jun is strong enough to be reproduced in vitro and constitutes the basis for the solid-phase kinase assay (**Fig. 1**; **ref.** *14*). The JNK in-gel kinase assay is based on the methodology described by Kameshita and Fujisawa, which enables the detection of kinase activity in sodium dodecyl sulfate (SDS) polyacrylamide gels *(15)*. Slight modifications to this method were introduced by Hibi et al. *(14)*. As mentioned above, JNKs are encoded by three genes, which give rise to four JNK1, four JNK2, and two JNK3 isoforms by alternative splicing. Because of this differential mRNA processing, 46-kDa and 55-kDa polypeptides are generated from each *jnk* gene *(7)*. In consequence, bands corresponding to both relative mobilities on SDS-polyacrylamide gel electrophoresis (PAGE) are usually detected in the in-gel kinase assay (**Fig. 1**).

2. Materials

1. Substrates GST-cJun and GST-ATF2 are commercially available or may be easily prepared if appropriate pGEX expression vectors are available, using the procedure described in **Subheading 3.1.**
2. Luria broth (LB): Make up in 1-L quantities. pH should be adjusted to 7.0 with NaOH, autoclaved, and stored at room temperature: 10 g/L tryptone, 5 g/L yeast extract, and 5 g/L NaCl.
3. Phosphate-buffered saline (PBS): May be prepared as a 10× stock solution, autoclaved, and stored at 4°C: 43 mM Na$_2$HPO$_4$, 14 mM KH$_2$PO$_4$, 1.37 M NaCl, and 27 mM KCl. Working solution is prepared by diluting 1:10 this stock, with its pH around 7.3, autoclaving, and storing at 4°C.
4. GST elution buffer: Make fresh daily: 50 mM N-hydroxyethylpiperazine-N'-2-ethanesulfonate (HEPES), pH 8.0, and 10 mM reduced glutathione.
5. JNK IP lysis buffer: Make up in 100-mL quantities and store at 4°C: 20 mM HEPES, pH 7.5, 10 mM ethylenebis(oxyethylenenitrilo)tetra-

acetic acid (EGTA), 2.5 mM MgCl$_2$, 40 mM β-glycerophosphate, and 1% Nonidet P-40. Before use, add 2 mM sodium orthovanadate, 1 mM dithiothreitol (DTT), 0.5 mM phenylmethyl sulfonyl fluoride (PMSF), 1 μg/mL aprotinin, and 1 μg/mL leupeptin.

6. JNK IP washing buffer: Make fresh daily, and ice-cold: PBS supplemented with 1% Nonidet P-40, and 2 mM sodium orthovanadate.

7. JNK buffer: Make up in 100-mL quantities and store at 4°C: 20 mM HEPES, pH 7.5, 20 mM MgCl$_2$, and 20 mM β-glycerophosphate. Before use, add 0.1 mM sodium orthovanadate and 2 mM DTT.

8. JNK reaction buffer: Make up in 10-mL quantities, and store as 1-mL aliquots at –20°C: 20 mM HEPES-Na, pH 7.5, 20 mM MgCl$_2$, 20 mM β-glycerophosphate, 0.1 mM sodium orthovanadate, and 2 mM DTT. Before use, add 20 μM ATP and enough [γ-^{32}P]ATP to yield 1 μCi/assay.

9. 4× Laemmli sample buffer: Make up in 10-mL quantities, and store as 1-mL aliquots at –20°C: 200 mM Tris-HCl, pH 6.8, 400 mM DTT, 8% SDS, 40% glycerol, and 0.2% bromophenol blue. 2× Laemmli sample buffer is prepared by adding 1 volume of water to this solution and is stored at –20°C.

10. WCE300 buffer: Make up in 100-mL quantities and store at 4°C: 25 mM HEPES, pH 7.7, 300 mM NaCl, 0.2 mM EDTA, 1.5 mM MgCl$_2$, 20 mM β-glycerophosphate, and 0.1% Triton X-100. Before use, add 0.1 mM sodium orthovanadate, 1 mM DTT, 0.5 mM PMSF, 1 μg/mL aprotinin, and 1 μg/mL leupeptin.

11. Dilution buffer: Make up in 100-mL quantities and store at 4°C: 20 mM HEPES, pH 7.7; 0.1 mM EDTA, 2.5 mM MgCl$_2$, 20 mM β-glycerophosphate, and 0.05% Triton X-100. Before use, add 0.1 mM sodium orthovanadate, 1 mM DTT, 0.5 mM PMSF, 1 μg/mL aprotinin, and 1 μg/mL leupeptin.

12. HEPES binding buffer: Make up in 100-mL quantities and store at 4°C: 20 mM HEPES, pH 7.7; 50 mM NaCl, 0.1 mM EDTA, 2.5 mM MgCl$_2$, 20 mM β-glycerophosphate, and 0.05% Triton X-100. Before use, add 0.1 mM sodium orthovanadate, 1 mM DTT, 0.5 mM PMSF, 1 μg/mL aprotinin, and 1 μg/mL leupeptin.

13. Isopropanol buffer: Make fresh daily: 50 mM HEPES, pH 7.6, and 20% isopropanol.

14. Buffer A: Make fresh daily: 50 mM HEPES, pH 7.6, and 5 mM 2-mercaptoethanol.

15. Denaturing buffer: Make fresh daily: 50 mM HEPES, pH 7.6, 5 mM 2-mercaptoethanol, and 6 M urea.

16. Renaturing buffer: Make fresh daily and ice-cold: 50 mM HEPES, pH 7.6, 5 mM 2-mercaptoethanol, and 0.05% Tween-20.

17. Washing buffer: Make up in 1-L quantities and store at room temperature: 5% trichloroacetic acid and 1% sodium pyrophosphate.

18. p38 IP lysis buffer: Make up in 100-mL quantities and store at 4°C: 20 mM Tris-HCl, pH 7.5, 137 mM NaCl, 2 mM EDTA, 2.5 mM MgCl$_2$, 50 mM β-glycerophosphate, 10% glycerol, and 1% Triton X-100. Before use, add 1 mM sodium orthovanadate, 1 mM DTT, 1 mM PMSF, 1 µg/mL aprotinin, and 1 µg/mL leupeptin.

19. p38 IP washing buffer: Make up in 100-mL quantities and store at 4°C: 100 mM Tris-HCl, pH 7.5, and 0.5 M LiCl.

20. p38 buffer: Make up in 100-mL quantities and store at 4°C: 25 mM Tris-HCl, pH 7.5, 1 mM EGTA, 25 mM MgCl$_2$, and 25 mM β-glycerophosphate. Before use add 0.5 mM sodium orthovanadate and 2 mM DTT.

21. p38 reaction buffer: Make up in 10-mL quantities and store as 1-mL aliquots at –20°C: 25 mM Tris-HCl, pH 7.5, 1 mM EGTA, 25 mM MgCl$_2$, 25 mM β-glycerophosphate, 0.5 mM sodium orthovanadate, and 2 mM DTT. Before use add 20 µM ATP, 0.1 µg/mL GST-ATF2 (*see* **Subheading 2.1.**), and enough [γ-^{32}P]ATP to yield 2.5 µCi/assay.

3. Methods

3.1. Preparation of Substrates

1. Transform *Escherichia coli* BL21 (DE3) strain with the appropriate pGEX expression vector, and select transformants growing up colonies in LB supplemented with 100 µg/mL ampicillin (LB-ampicillin).

2, Inoculate a colony in 25 mL of LB-ampicillin and grow overnight at 37°C in a shaker.

3. Dilute this culture 1:10 in LB-ampicillin and continue to grow for 1 h at 37°C in a shaker.

4. Add IPTG to a final concentration of 100 µM and incubate at 37°C in a shaker for 3 h.

5. Collect bacteria by centrifugation at 5000g for 10 min at room temperature and discard supernatant.

6. Suspend pellet in 10 mL of ice-cold PBS and lyse bacteria by sonication.

7. Add 1 mL of a 10% Triton X-100 solution, mix, and clear the lysate by centrifugation at 10,000g for 10 min at 4°C.

8. Collect supernatant, mix it with 0.5 mL of 50% glutathione-agarose beads (*see* **Note 1**) slurry, and rotate for 1 h at 4°C.
9. Centrifuge for 1 min at 500*g* at 4°C and discard supernatant.
10. Suspend beads in 10 mL of ice-cold PBS, centrifuge 1 min at 500*g* at 4°C, and discard supernatant. Repeat this step twice more.
11. Suspend beads in 250 μL of GST elution buffer, transfer to a microfuge tube and rotate for 20 min at 4°C.
12. Centrifuge for 10 s at 12,000*g*, collect supernatant, and dialyze it overnight at 4°C against PBS containing 10% glycerol.
13. Store in small aliquots at −70°C (*see* **Note 2**).

3.2. JNK Immunocomplex Assay

1. Harvest cells. For adherent cells, aspirate medium, and wash cell monolayer once with PBS. Add 1 mL of PBS, scrape cells with a rubber policeman, transfer into a microfuge tube, and spin at 3000*g* for 2 min. For cells growing in suspension, collect cells by centrifugation at 1000*g* in a tabletop centrifuge for 5 min, aspirate medium, suspend them in 1 mL of PBS, transfer into a microfuge tube, and spin at 3000*g* for 2 min. From this step and unless indicated, samples should be kept on ice, centrifugation steps run at 4°C, and buffers should be ice-cold.
2. Aspirate PBS and suspend pellet in JNK IP lysis buffer (500 μL for 5×10^6 cells).
3. Rotate cell extract for 20 min, clear it by centrifugation at 12,000*g* for 10 min, and transfer supernatant into a new microfuge tube.
4. Measure protein concentration by any standard method such as Bradford.
5. Transfer 100 μg of the cell extract (*see* **Note 3**) into a microfuge tube containing 20 μL of a 50% slurry of protein A-Sepharose beads (*see* **Note 1**) and 1 μL of anti-JNK antibody (*see* **Note 4**). Bring immunoprecipitation mixture to a final volume of 500 μL with JNK IP lysis buffer and rotate for 2 h.
6. Wash immunocomplexes by adding 1 mL of JNK IP washing buffer, mixing, and spinning for 10 s at 12,000*g*. Repeat this step twice more and perform a final wash with 1 mL of JNK buffer.
7. Suspend gently immunocomplexes in 30 μL of JNK reaction buffer supplemented with 1 μg of the substrate GST-cJun, and incubate for 20 min at 30°C.

8. Stop reaction by adding 15 µL of 4× Laemmli sample buffer. Incubate at 100°C for 3 min and load onto a 12% SDS-PAGE.

9. After electrophoresis, fix the gel in isopropanol:water:acetic acid (25:65:10; gel staining is optional), dry, and autoradiograph.

3.3. JNK Solid-Phase Kinase Assay

1. Harvest cells as described in **Subheading 3.2., step 1**. From this step and unless indicated, samples should be kept on ice, centrifugation steps run at 4°C, and buffers should be ice-cold.

2. Aspirate PBS and suspend pellet in WCE300 buffer (200 µL for 5×10^6 cells).

3. Rotate cell extract for 30 min, clear by centrifugation at 12,000g for 10 min, and transfer supernatant into a new microfuge tube.

4. Dilute the extract with three volumes of dilution buffer, place on ice for 10 min, and clear by centrifugation at 12,000g for 10 min.

5. Transfer supernatant into a new microfuge tube and determine protein concentration by any standard method such as Bradford.

6. Transfer 500 µg of the cell extract into a microfuge tube containing 10 µg of GST-cJun 20 µL of 50% slurry of glutathione-agarose beads (*see* **Note 1**) and rotate for 3 to 5 h.

7. Wash beads by adding 1 mL of HEPES binding buffer, mixing, and spinning for 10 s at 12,000g. Repeat this step four more times.

8. Suspend gently the beads in 30 µL of JNK reaction buffer and incubate for 20 min at 30°C.

9. Stop reaction by adding 1 mL of HEPES binding buffer, mixing, and spinning for 10 s at 12,000g.

10. Discard supernatant and suspend beads in 30 µL of 2× Laemmli sample buffer. Incubate at 100°C for 3 min and load onto a 12% SDS-PAGE.

11. After electrophoresis, fix the gel in isopropanol:water:acetic acid (25:65:10; gel staining is optional), dry, and autoradiograph.

3.4. JNK In-Gel Kinase Assay

1. Prepare the substrate GST-cJun as described in **Subheading 3.1.** up to **step 10**. Elute protein by mixing the beads with 650 µL of a 0.75% SDS solution, transfer to a microfuge tube, centrifuge 10 for s at 12,000g, and use the supernatant to prepare the separating gel of a

10% SDS-polyacrylamide gel (*see* **Note 5**). The stacking gel should be prepared as a regular one.

2. Mix 1 volume of cell extract containing 20 μg of protein with the same volume of 2× Laemmli sample buffer, incubate at 100°C for 3 min, and load onto the 10% SDS-PAGE prepared as described above.
3. After electrophoresis, perform 15-min washes of the gel at room temperature in 100 mL of the following solutions: twice in isopropanol buffer, twice in buffer A, and twice in denaturing buffer.
4. Perform 15-min washes of the gel at 4°C in 100 mL of renaturing buffer containing either 3, 1.5, or 0.75 *M* urea.
5. Wash the gel for 15 min in 150 mL of renaturing buffer at 4°C. Repeat this step three more times.
6. Wash the gel for 30 min in 50 mL of JNK buffer at 4°C.
7. Incubate the gel for 1 h in 15 mL of JNK buffer containing 20 μ*M* ATP and 5 μCi/mL of [γ-^{32}P]ATP at 30°C.
8. Wash the gel in 200 mL of washing buffer at room temperature for 30 min. Repeat this step four times more.
9. Dry the gel and autoradiograph.

3.5. p38 Immunocomplex Assay

1. Harvest cells as described in **Subheading 3.2., step 1**. From this step and unless indicated, samples should be kept on ice, centrifugation steps run at 4°C, and buffers should be ice-cold.
2. Aspirate PBS and suspend pellet in p38 IP lysis buffer (500 μL for 5 × 10^6 cells).
3. Rotate cell extract for 20 min, clear it by centrifugation at 12,000*g* for 10 min, and transfer supernatant into a new microfuge tube.
4. Measure protein concentration by any standard method such as Bradford.
5. Transfer 200 μg of the cell extract into a microfuge tube containing 20 μL of a 50% slurry of protein A-Sepharose beads (*see* **Note 1**) and 1 μL of anti-p38 antibody (*see* **Note 6**). Bring immunoprecipitation mixture to a final volume of 500 μL with p38 IP lysis buffer and rotate for 2 h.
6. Wash immunocomplexes by adding 1 mL of p38 IP lysis buffer, mixing and spinning for 10 s at 12,000*g*. Repeat this step and perform two additional 1-mL washes of the immunocomplexes with p38 IP washing buffer and p38 buffer.

7. Suspend gently immunocomplexes in 30 µL of p38 reaction buffer and incubate for 20 min at 30°C.
8. Stop reaction by adding 15 µL of 4× Laemmli sample buffer. Incubate at 100°C for 3 min and load onto a 12% SDS-PAGE.
9. After electrophoresis, fix the gel in isopropanol:water:acetic acid (25:65:10; gel staining is optional), dry, and autoradiograph.

4. Notes

1. Both glutathione-agarose and protein A-Sepharose beads may be purchased as lyophilized powder stabilized with carbohydrates. Before use, they need to be preswollen in distilled water for 1 h, washed extensively, and equilibrated with either PBS or the appropriate buffer. Alternatively, they may be purchased already preswollen. In this case, they only need to be equilibrated with the appropriate buffer by several washes. They are stable for up to 1 mo if stored as a 50% (v/v) slurry at 4°C.
2. Protein concentration may be determined by any standard method; generally, it is >1 µg/µL. Protein preparation may be analyzed in 12% SDS-PAGE. Most frequent GST-cJun and GST-ATF2 recombinant proteins used as substrates contain amino acids 1 to 79 and 1 to 96 of c-Jun and ATF2, respectively. Therefore, the size of these GST-fusion proteins is approx 35 kDa. Repeated thawing and freezing should be avoided.
3. The amount of cell extract needed to determine JNK activity may vary among different cell lines or transfection efficiency (*see* **Note 4**). The protocol outlined has been standardized to assay endogenous JNK activity from HeLa cells, although in our hands it has been successful with many other cell lines as well as mouse brain extracts. In addition, this protocol is also suitable for assaying JNK activity in transient transfection experiments, where it is overexpressed as either hemagglutinin (HA) or Flag-tagged JNK (*see* **Note 4**).
4. Antibodies to immunoprecipitate JNK as an active kinase are available from various commercial sources, and so the amount of antibody required may vary. Routinely, we use the one supplied by Santa Cruz (sc-474). In transient transfection experiments where JNK is overexpressed as a fusion protein with either HA or Flag epitope, anti-HA (12CA5) or anti-Flag (M2) antibodies should be used. These antibodies are available from various commercial sources.

5. The final concentration of GST-cJun in the gel should be around 50 µg/mL.
6. Antibodies to immunoprecipitate p38 as an active kinase are available from various commercial sources, and so the amount of antibody required may vary. Routinely, we use the one supplied by Santa Cruz (sc-535). In transient transfection experiments where p38 is overexpressed as a fusion protein with either HA or Flag epitope, anti-HA (12CA5) or anti-Flag (M2) antibodies should be used. These antibodies are available from various commercial sources.

Acknowledgments

This study was funded by grants from the Plan General del Conocimiento (PM97-0115) and Plan Nacional de Investigación y Desarrollo (1FD97-0281-CO2-01) from the Ministerio de Educación y Cultura of Spain, and Fundació "La Caixa" (99/032-00). M.M. was supported by a fellowship from the Ministerio de Educación y Cultura of Spain.

References

1. Robinson, M. J. and Cobb, M. H. (1997) Mitogen-activated protein kinase pathways. *Curr. Opin. Cell. Biol.* **9,** 180–186.
2. Ip, Y. T. and Davis, R. J. (1998) Signal transduction by the c-Jun N-terminal kinase (JNK)—from inflammation to development. *Curr. Opin. Cell. Biol.* **10,** 205–219.
3. Ichijo, H. (1999) From receptors to stress-activated MAP kinases. *Oncogene* **18,** 6087–6093.
4. Leppä, S. and Bohmann, D. (1999) Diverse functions of JNK signaling and c-Jun in stress response and apoptosis. *Oncogene* **18,** 6158–6162.
5. Xia, Z., Dickens, M., Raingeaud, J., and Davis, R. J. (1995) Opposing effects of ERK and JNK-p38 MAP kinases on apoptosis. *Science* **270,** 1326–1331.
6. Lenczowski, J. M., Dominguez, L., Eder, A. M., King, L. B., Zacharchuk, C. M., and Ashwell, J. D. (1997) Lack of a role for Jun kinase and AP-1 in Fas-induced apoptosis. *Mol. Cell. Biol.* **17,** 170–181.
7. Gupta, S., Barret, T., Whitmarsh, A. J., Cavanagh, J., Sluss, H. K., Dérijard, B., and Davis, R. J. (1996) Selective interaction of JNK

protein kinase isoforms with the transcription factors. *EMBO J.* **15,** 2760–2770.

8. Yang, D., Kuan, C. Y., Whitmarsh, A. J., Rincon, M., Zheng, T. S., Davis, R. J., Rakic, P., and Flavell, R. A. (1997) Absence of excitotoxicity-induced apoptosis in the hippocampus of mice lacking the JNK3 gene. *Nature* **389,** 865–870.

9. Tournier, C, Hess, P., Yang, D. D., Xu, J., Turner, T. K., Nimnual, A., et al. (2000) Requirement of JNK for stress-induced activation of the cytochrome c mediated death pathway. *Science* **288,** 870–874.

10. Kuan, C. Y., Yang, D. D., Samanta Roy, D. R., Davis, R. J., Rakic, P., and Flavell, R. A. (1999) The Jnk1 and Jnk2 protein kinases are required for regional specific apoptosis during early brain development. *Neuron* **22,** 667–676.

11. Nishina, H., Fischer, K. D., Radvanyl, L., Shahinian, A., Hakem, R., Ruble, E. A., et al. (1997) Stress-signaling kinase Sek1 protects thymocytes from apoptosis mediated by CD95 and CD3. *Nature* **385,** 350–353.

12. Nebreda, A. R. and Porras, A. (2000) p38 MAP kinases: beyond the stress response. *Trends Biochem. Sci.* **25,** 257–260.

13. Dérijard, B., Hibi, M., Wu, I. H., Barret, T., Su, B., Deng, T., Karin, M., and Davis, R. J. (1994) JNK1: a protein kinase stimulated by UV light and Ha-Ras that binds and phosphorylates the c-Jun activation domain. *Cell* **76,** 1025–1037.

14. Hibi, M., Lin, A., Smeal, T., Minden, A., and Karin, M. (1993) Identification of an oncoprotein- and UV-responsive protein kinase that binds and potentiates the c-Jun activation domain. *Genes Devel.* **7,** 2135–2148.

15. Kameshita, I. and Fujisawa, H. (1989) A sensitive method for detection of calmodulin-dependent protein kinase II activity in sodium dodecyl sulfate-polyacrylamide gel. *Anal. Biochem.* **183,** 139–143.

11

Methods for Culturing Primary Sympathetic Neurons and for Determining Neuronal Viability

Jonathan Whitfield, Stephen J. Neame, and Jonathan Ham

Summary

Developing nerve growth factor (NGF)-dependent sympathetic neurons are one of the best-studied in vitro models of neuronal apoptosis and have been used to identify key components of the neuronal cell death pathway. This chapter describes how to prepare purified cultures of primary sympathetic neurons and how to induce apoptosis by NGF deprivation. In addition, a simple method for measuring neuronal viability based on the live/dead assay is also described. This can be used for assessing the effect of small molecule inhibitors of protein kinases, caspases and other enzymes, on NGF withdrawal-induced death.

Key Words: Sympathetic neuron; nerve growth factor; developmental neuronal death; apoptosis; live/dead cell viability assay.

1. Introduction

This chapter and the following chapter describe methods for detecting apoptosis in primary sympathetic neurons cultured in vitro, but the same techniques can be applied to other neuronal types, such as cerebellar granule neurons or cortical neurons. The mechanisms of neuronal apoptosis are of considerable interest because this form of cell death plays an important role in the development of the mammalian nervous system and because there is increasing evidence that neurons die by apoptosis in human neurodegenerative

From: *Methods in Molecular Biology, vol. 282: Apoptosis Methods and Protocols*
Edited by: H. J. M. Brady © Humana Press Inc., Totowa, NJ

diseases *(1)*. Sympathetic neurons, which form part of the periph-eral nervous system, have proved to be a particularly useful model system for in vitro studies of neuronal death. During mammalian development one third of these cells normally die by apoptosis dur-ing the first 2 wk after birth and at this time they require nerve growth factor (NGF) for survival *(2)*. Sympathetic neurons can be isolated from 1-d-old rats and cultured in vitro for extended periods in the presence of NGF. If NGF is removed from the growth medium, sympathetic neurons undergo a form of cell death that has the classic features of apoptosis: rates of RNA and protein synthesis fall, the cytoplasm and nucleus shrink, mitochondria release cyto-chrome *c* into the cytoplasm, caspases are activated and required for cell death, chromatin condenses and the DNA is degraded into oligonucleosomal fragments, phosphatidylserine redistributes from the inner to the outer surface of the plasma membrane, neurites frag-ment, and the cells detach from their substrate (**Fig. 1**; **refs. *3–11***).

A major advantage of the sympathetic neuron model is that its cell biology is very well defined in vitro and in vivo. In addition, because it is a primary neuron model of apoptosis, it has advantages over neu-ronal cell lines, such as PC12 cells, which were derived from a tumour (a rat phaeochromocytoma) and which contain a number of unde-fined genetic changes. However, a significant limitation of the sym-pathetic neuron system is that relatively small numbers of neurons are obtained in a typical preparation (100,000–125,000 neurons from 10 1-d-old rats), which makes it harder to conduct biochemistry and molecular biology experiments. For this reason, methods for study-ing apoptosis that work at the single cell level or that only require small amounts of protein or RNA are advantageous. In this chapter, we will describe the following techniques: the preparation of puri-fied sympathetic neuron cultures and the induction of apoptosis by NGF withdrawal and the application of the live/dead cell viability assay to unfixed living cells. In the following chapter (*see* Chapter 12), immunocytochemical techniques for studying apoptosis at the single neuron level will be described.

The method for preparing sympathetic neurons depends on the following principles. The superior cervical ganglia (SCG) of 1-d-old

+ NGF   - NGF

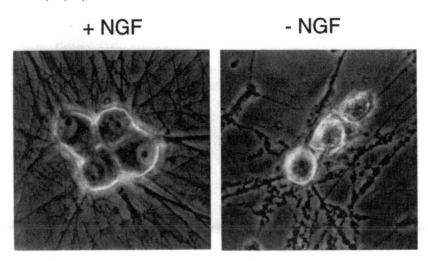

Fig. 1. Morphology of sympathetic neurons grown in SCG medium containing NGF or after 48 h of NGF deprivation. Sympathetic neurons were cultured on glass coverslips in the presence of NGF for 6 d and then refed with fresh SCG medium containing NGF (+NGF) or lacking NGF and supplemented with a neutralizing anti-NGF antibody (–NGF) as indicated. Forty-eight hours later, the cells were examined on a Zeiss Axiovert S100 microscope. Photographs were taken on phase contrast and 40 times magnification using Kodak TMY 400 black and white film. In the presence of NGF (+NGF) the nuclei and nucleoli are clearly visible and the neurons have smooth cell bodies and an extensive network of neurites. At 48 h after NGF withdrawal (–NGF), many of the neurites have disintegrated and the cell bodies are considerably smaller. The nuclei can no longer be seen clearly. The bar represents 25 μM.

rats are used as a source of sympathetic neurons. After isolation, the ganglia are dissociated by enzymatic treatment with trypsin and collagenase followed by mechanical dissociation by trituration. In addition to sympathetic neurons, SCGs contain Schwann cells and fibroblasts. However, the cell suspension can be enriched for sympathetic neurons by plating the cells in an uncoated tissue culture dish, to which Schwann cells and fibroblasts can adhere but neurons cannot (a preplating step). The neurons left in suspension are then plated in NGF-containing medium on glass coverslips or tissue culture plastic coated with poly-L-lysine and laminin, and antimitotic agents are added to the culture medium to inhibit the proliferation of non-neuronal cells. After 5–7 d in culture, the neurons will have grown in size and formed an extensive network of neurites and can be used for cell death experiments. Apoptosis is induced by replacing the culture medium with medium-lacking NGF that has been supplemented with a neutralizing anti-NGF antibody. Most of the sympathetic neurons are dead by 3 d after NGF withdrawal.

The percentage of viable cells in an unfixed population of sympathetic neurons can be easily determined by conducting a live/dead assay using a kit developed by Molecular Probes, Inc. In this assay two dyes, calcein AM and ethidium homodimer 1, are added directly to the culture medium. Viable cells convert calcein AM to calcein, which fluoresces green, and exclude ethidium homodimer 1. However, apoptotic cells that have lost membrane integrity take up ethidium homodimer, which binds to DNA, and consequently have red nuclei. After staining, the cells are examined on an inverted fluorescence microscope, and the percentage of live cells is determined (**Fig. 2**). This is a convenient assay to use when testing the effect of growth factors and small molecule inhibitors of protein kinases or caspases on sympathetic neuron survival in the presence or absence of NGF.

2. Materials

2.1. Sympathetic Neuron Prep

1. L15+ medium: L15+ is L15 (Leibovitz) medium with L-glutamine and L-amino acids (Life Technologies Ltd., 11415-049) supplemented with 0.1% fetal calf serum and penicillin/streptomycin (Life Technologies Ltd., 15140-122).
2. Trypsin solution: Trypsin (Sigma, T-4665) is dissolved at 0.025% in phosphate-buffered saline (PBS) without calcium chloride and magnesium chloride (Sigma, D-8537). The stock is stored as 5-mL aliquots at –20°C.
3. Collagenase solution: Collagenase type 2 is purchased from the Worthington Biochemical Corporation and stored dessicated at 4°C.

calcein AM   EthD-1

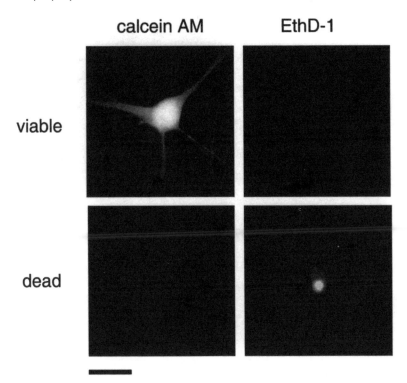

Fig. 2. Typical staining pattern and morphology of viable and dead sympathetic neurons in a live/dead cell viability/cytotoxicity assay. Sympathetic neurons were cultured on glass coverslips in the presence of NGF for 6 d and then refed with fresh SCG medium containing NGF or lacking NGF and supplemented with neutralizing anti-NGF antibody. After 48 h, the cells were stained with calcein AM and ethidium homodimer 1 (EthD-1) as described in **Subheading 3.2.** Viable cells stain brightly with calcein AM and exclude ethidium homodimer. Apoptotic neurons that have lost membrane integrity do not stain with calcein and no longer exclude ethidium homodimer. The latter stains the pyknotic nuclei. The bar represents 25 μM.

Just before use, the collagenase is dissolved at 0.2% in PBS with calcium chloride and magnesium chloride (Sigma, D-8662) and sterilized by filtration. The collagenase should be batch tested (*see* **Note 1**).
4. SCG medium: SCG medium is DMEM containing glucose at 4.5 g/L and sodium pyruvate at 110 mg/L (Sigma, D-6546) supplemented with 10% fetal calf serum, penicillin/streptomycin, and L-glutamine at 2 mM (Life Technologies, 25030-024).

5. Antimitotic agents: 20 m*M* (1000×) stock solutions of uridine (Sigma U-3003) and fluorodeoxyuridine (Sigma, F-0503) are prepared in sterile water, filter sterilized, and stored as small aliquots at –20°C. When required, these are added to the SCG medium just before use at a final concentration of 20 µ*M*. If necessary, aphidicolin can also be added to 3.3 µg/mL (*see* **Note 3**). Aphidicolin (Sigma A-0781) is dissolved in DMSO at 10 mg/mL and can be stored for up to 6 wk at –20°C.

6. NGF: Lyophilized, ultrapure 2.5S Nerve Growth Factor is purchased from Cedarlane Laboratories, Ltd. (CLMCNET-001.25) and reconstituted according to the manufacturer's instructions at 50 µg/mL in PBS. This stock solution is aliquoted and stored at –80°C.

7. Glass coverslips coated with poly-L-lysine and laminin: BDH type 1 coverslips (circular, 13-mm diameter) are coated with poly-L-lysine as follows. First, the coverslips are baked in glass Petri dishes at 140°C for 4 h. The baked coverslips can be stored in a sealed plastic sandwich box containing silica gel. Poly-L-lysine (Sigma, P-1524) is dissolved in sterile, tissue culture-quality water at 1 mg/mL. The baked coverslips are placed in a 9-cm diameter tissue culture dish and covered with poly-L-lysine solution. The sealed dish is then placed on a shaker and gently rotated for 24 h at room temperature. After coating, the poly-L-lysine solution is removed (but stored at –20°C; it can be re-used up to three times) and the coverslips are washed with tissue culture water overnight for 3 d, changing the water each day. The coverslips are dried by standing them around the edge of a dry tissue culture dish inside a class II tissue culture hood (usually overnight) and are then stored at room temperature in a tissue culture dish placed inside a sealed plastic sandwich box containing silica crystals. The poly-L-lysine-coated coverslips are coated with laminin as follows. Laminin (1 mg/mL) is purchased from Sigma (L-2020), thawed on ice, and divided into 20-µL aliquots in 1.5-mL microfuge tubes, which are stored at –80°C. On the day of a sympathetic neuron prep, aliquots of laminin are thawed on ice and then diluted to 20 µg/mL in room temperature DMEM. Individual poly-L-lysine-coated coverslips are placed in the centre of 3.5-cm diameter tissue culture dishes and 100 µL of the diluted laminin solution is pipetted onto each coverslip so as to cover the whole surface. If fewer neurons are to be plated, a smaller volume of laminin should be used (*see* Notes section). The coverslips are left at room temperature for 2 h. Just before plating the cells, the laminin solution is aspirated off.

8. Anti-NGF antibody: Lyophilized anti-NGF monoclonal antibody (Chemicon MAB5260-GOUG) is dissolved in PBS at 25 µg/mL. This solution can be aliquoted and stored at –20°C. Once thawed, the antibody solution should be stored at 4°C and added to the medium in NGF withdrawal experiments at a final concentration of 100 ng/mL.

2.2. Live/Dead Assay

The live/dead viability/cytotoxicity kit from Molecular Probes (L-3224) is used. This contains stock solutions of calcein AM (4 mM) and ethidium homodimer 1 (2 mM), which are stored at –20°C.

1. 10× working solution of calcein AM and ethidium homodimer 1: just before use, mix 800 µL of prewarmed DMEM, 1 µL of 4 mM calcein AM, and 4 µL of ethidium homodimer 1 in a 1.5-mL microfuge tube.

3. Methods

3.1. Isolation and Culture of Sympathetic Neurons

All steps are conducted at room temperature unless stated otherwise. Dissection tools should be sterilized beforehand by soaking them in 70% ethanol. Dissection is performed using a stereo dissection microscope and light source, for example, Leica MZ 7.5 microscope and Intralux 4000-1 light source (Volpi AG, Switzerland), placed inside a horizontal laminar flow hood.

1. Dissect the superior cervical ganglia from 1-d-old Sprague–Dawley rats and place in prewarmed L15+ medium in a 9-cm diameter tissue culture dish. The preganglionic nerve fiber that leads to the superior cervical ganglion (the cervical sympathetic trunk) runs parallel to the carotid artery and the superior cervical ganglion is found at the point where the carotid artery bifurcates into the internal and external carotid arteries. It is important to distinguish the SCG from the nodose ganglion, which is also near the carotid branch point. The nodose ganglion is smaller than the SCG and is connected to a thicker nerve fiber.
2. Use sterilized Dumont no. 5 forceps (Agar Scientific Limited) to separate the ganglia from extraneous tissue and to remove any

ensheathing fibroblasts and blood vessels. Also remove the nerve
fibers that are attached to each end of the ganglion.

3. Use a pair of sterilized fine forceps to transfer the cleaned ganglia to
a 15-mL Falcon tube containing 5 mL of prewarmed 0.025% trypsin
solution. Incubate at 37°C for 30 min.

4. Carefully remove the trypsin, but not the ganglia, and add 5 mL of
0.1% collagenase solution. Incubate at 37°C for 30 min.

5. Add 10 mL of SCG medium containing NGF (50 ng/mL), fluorodeoxyuridine (20 μM), and uridine (20 μM) and then remove the supernatant and transfer it to a 50-mL Falcon tube. Add 1 mL of SCG
medium + NGF, FU, and U to the ganglia and gently triturate using a
P1000 Gilson pipetman and blue tip until the ganglia are fully dis-
persed. Avoid air bubbles. Check 20 μL of the suspension under the
microscope to make sure that the cells are not in clumps or overtriturated (*see* **Note 2**).

6. Pool the cell supernatant (15 mL) and the cell suspension (1 mL) in
the 50-mL Falcon tube and centrifuge at 200*g* for 10 min at room
temperature in a bench centrifuge. Carefully remove the supernatant
and gently resuspend the cell pellet in 1 mL of SCG medium + NGF,
FU, and U using a blue tip. Again, check under the microscope how
well the cells are separated. Then make the volume up to 10 mL with
SCG medium + NGF, FU, and U (*see* **Note 3**).

7. Pipet the cell suspension into an uncoated 9-cm diameter tissue culture dish. Incubate at 37°C in 10% CO_2 for 2–3 h (preplating step).

8. After preplating, look at the dish under the microscope. Non-
neuronal cells, which are phase dark, will have stuck to the dish and
have started to flatten out. The neurons are phase bright and will be
in suspension or loosely attached to the dish. After the preplating,
knock the dish twice and then transfer the cell suspension to a 50-mL
Falcon tube.

9. Spin the cells at 200*g* for 10 min at room temperature. Carefully
remove the supernatant and add up to 1 mL of medium. Gently resuspend the cells using a blue tip.

10. Mix 20 μL of cell suspension with 20 μL of 0.4 % trypan blue solution (Sigma, T-8154) and count in a 0.2-mm deep, modified Fuchs-
Rosenthal hemocytometer. Count the number of neurons in 16 large
squares and calculate the number per milliliter. No./mL = no. in 16
squares × 312.5 × 2. 10,000–12,500 neurons are usually obtained
from one pair of ganglia (one rat pup; *see* **Note 4**).

11. Plate the neurons on coverslips placed in 3.5-cm tissue culture dishes. Plate 8,000–10,000 neurons in 100 µL of SCG medium + NGF, FU, and U per coverslip. Fewer cells can be plated in a smaller area if required (*see* **Note 5**). Incubate for 2 h at 37°C in 10% CO_2 to allow the neurons time to stick down. Then flood each dish with 2 mL of medium. The neurons can also be plated in 3.5-cm diameter tissue culture dishes coated with poly-L-lysine and laminin. 5.5×10^4 neurons are plated in 2 mL of medium per 3.5-cm dish.

12. Refeed the neurons with fresh SCG medium + NGF, FU, and U at 1 and 2 d after plating and then every 3 d. After 5–7 d in vitro, apoptosis can be induced by depriving the neurons of NGF (*see* **Note 6**). The medium is gently removed from each dish and replaced with SCG medium containing FU and U but lacking NGF (a rinse step). This is then removed and replaced with SCG medium lacking NGF supplemented with FU, U, and anti-NGF antibody at 100 ng/ mL. The medium changes need to be conducted very carefully so as to avoid damaging the network of neurites that attach the neurons to their substrate. The cells are then returned to a 37°C/10% CO_2 incubator. At 3 d after NGF withdrawal, most of the neurons should be dead (*see* **Fig. 1**).

3.2 Live/Dead Cell Viability Assay

To conduct a live/dead assay, grow the sympathetic neurons on 13-mm diameter glass coverslips placed in 3.5-cm tissue culture dishes in 2 mL of medium and use the live/dead assay kit from Molecular Probes (L-3224; *see* **Note 7**).

1. Prepare a 10× working solution of calcein AM and ethidium homodimer in prewarmed DMEM as described in **Subheading 2.2.**
2. Gently add 0.22 mL of the 10× working solution to the cells in 2 mL of medium.
3. Incubate at 37°C in 10% CO_2 for 30 min.
4. Count on an inverted fluorescence microscope, for example, a Zeiss Axiovert 100, on 20× magnification. Count 10 fields, starting at the top edge and tracking down the coverslip. In each field, count the number of live cells and the number of dead cells. Then, determine the total number of live and total number of dead cells in 10 fields. Live sympathetic neurons are evenly stained green and never stain red. Dead neurons always stain red but may also sometimes stain a speckly green

(*see* **Fig. 2**). Try not to leave the neurons more than an hour after the end of the 30-min incubation with the dyes otherwise the background becomes too high.

$$\% \text{ of viable cells} = \frac{\text{no. of live cells}}{\left(\text{No. of live cells} + \text{no. of dead cells}\right)} \times 100$$

4. Notes

1. Sympathetic neuron prep: it is important to test different batches of collagenase and serum to find out which batches are optimal for dissociation of the ganglia and for growth and survival of the neurons. When optimizing culture conditions, the morphology of the neurons after 7 d in vitro should be studied carefully using a phase contrast microscope. After 7 d, the neurons should have round nuclei with clear nucleoli and smooth cell bodies and will have formed an extensive network of neurites (**Fig. 1**, +NGF). If the cell bodies contain vacuoles or if the neurites have a beaded appearance the neurons are unhealthy.

2. The purity of sympathetic neuron cultures depends on how well the ganglia are cleaned up and the efficiency of the trituration step. If the cell suspension is insufficiently triturated the cells will remain as clumps containing both neurons and non-neuronal cells. If trituration through a P1000 tip does not separate neurons in clumps, then the cell suspension can be triturated further using a sterile 1-mL syringe and 19-gage needle or even 21- or 23-gage needles. However, if the cell suspension is overtriturated the cells will be damaged. Trituration is the most critical step in the whole procedure. If it has worked well, the cell suspension will mainly contain single whole cells, which often have short neurites.

3. If non-neuronal cells are a problem, the preplating step should be carried out for 3 h and aphidicolin can be added to the medium (at 3.3 µg/mL) in addition to the fluorodeoxyuridine and uridine.

4. Newborn or 2-d-old rats can be used instead of 1-d-old rats if the latter are not available.

5. Fewer neurons can be plated on each coverslip if a smaller area is coated with laminin (this keeps the cell density the same). For example, 3000–5000 cells can be plated in a volume of 40 µL. For some applications, such as single-cell microinjection experiments, only a few hundred neurons are required on each coverslip.

6. For NGF withdrawal experiments it is best to use the neurons between 6 and 8 d after plating. If sympathetic neurons are cultured in vitro for 3 to 4 wk they lose their dependence on NGF for survival. This also occurs in vivo during postnatal development.

7. Live/dead assay: the neurons should not be plated too densely or they will grow in large clumps that make counting difficult. It is important to try and plate neurons at the same density on each coverslip. +NGF and –NGF controls should always be performed. Each experiment should be scored in a blinded manner and performed three times to obtain statistically significant data. The average ± the standard error should be calculated for each treatment.

Acknowledgments

This work was supported by the Eisai Company of Japan and the Wellcome Trust. J.H. is a Wellcome Trust Senior Research Fellow in Basic Biomedical Science.

References

1. Yuan, J. and Yankner, B. A. (2000) Apoptosis in the nervous system. *Nature* **407,** 802–809.

2. Wright, L. L., Cunningham, T. J., and Smolen, A. J. (1983) Developmental neuron death in the rat superior cervical sympathetic ganglion: cell counts and ultrastructure. *J. Neurocytol.* **12,** 727–738.

3. Martin, D. P., Schmidt, R. E., DiStefano, P. S., Lowry, O. H., Carter, J. G., and Johnson, E. M. (1988) Inhibitors of protein synthesis and RNA synthesis prevent neuronal Death caused by nerve growth factor deprivation. *J. Cell. Biol.* **106,** 829–844.

4. Deckwerth, T. L., and Johnson, E. M. (1993) Temporal analysis of events associated with programmed cell death (apoptosis) of sympathetic neurons deprived of nerve growth factor. *J. Cell. Biol.* **123,** 1207–1222.

5. Edwards, S. N. and Tolkovsky, A. M. (1994) Characterization of apoptosis in cultured rat sympathetic neurons after nerve growth factor withdrawal. *J. Cell. Biol.* **124,** 537–546.

6. Rimon, G., Bazenet, C. E., Philpott, K. L., and Rubin, L. L. (1997) Increased surface phosphatidylserine is an early marker of neuronal apoptosis. *J. Neurosci. Res.* **48,** 563–570.

7. Martinou, I., Fernandez, P.-A., Missotten, M., White, E., Allet, B., Sadoul, R., et al (1995) Viral proteins E1B19K and p35 protect sympathetic neurons from cell death induced by NGF deprivation. *J. Cell. Biol.* **128,** 201–208.

8. Deshmukh, M., Vasilakos, J., Deckwerth, T. L., Lampe, P. A., Shivers, B. D., and Johnson, E. M. (1996) Genetic and metabolic status of NGF-deprived sympathetic neurons saved by an inhibitor of ICE family proteases. *J. Cell. Biol.* **135,** 1341–1354.

9. McCarthy, M. J., Rubin, L. L., and Philpott, K. L. (1997) Involvement of caspases in sympathetic neuron apoptosis. *J. Cell. Sci.* **110,** 2165–2173.

10. Neame, S. J., Rubin, L. L., and Philpott, K. L. (1998) Blocking cytochrome c activity within intact neurons inhibits apoptosis. *J. Cell. Biol.* **142,** 1583–1593.

11. Deshmukh, M. and Johnson, E. M. (1998) Evidence of a novel event during neuronal death: development of competence-to-die in response to cytoplasmic cytochrome c. *Neuron* **21,** 695–705.

12

Immunocytochemical Techniques for Studying Apoptosis in Primary Sympathetic Neurons

Stephen J. Neame, Jonathan Whitfield, and Jonathan Ham

Summary

Developing sympathetic neurons, which depend on nerve growth factor for survival, are one of the best studied in vitro models of neuronal apoptosis and have been extensively used for cellular and molecular studies of the neuronal death pathway. Important apoptotic events after nerve growth factor withdrawal include the release of proapoptotic proteins, such as cytochrome c, from the mitochondria and the activation of caspases, followed by nuclear DNA fragmentation and chromatin condensation. In this chapter, we describe immunocytochemical techniques for studying apoptotic DNA fragmentation, changes in nuclear morphology, and mitochondrial cytochrome c release at the single cell level using sympathetic neurons cultured on glass coverslips.

Key Words: Sympathetic neuron; nerve growth factor; apoptosis; TUNEL analysis; cytochrome c; immunocytochemistry.

1. Introduction

This chapter describes some of the techniques that can be used to study apoptosis in primary sympathetic neuron cultures at the single cell level, namely: 1) the detection of apoptotic DNA fragmentation and chromatin condensation by terminal deoxynucleotidyl transferase-mediated dUTP nick end labeling (TUNEL) analysis and bisbenzimide (Hoechst 33342) staining; and 2) the visualization of

From: *Methods in Molecular Biology, vol. 282: Apoptosis Methods and Protocols*
Edited by: H. J. M. Brady © Humana Press Inc., Totowa, NJ

alterations in cytochrome c distribution that occur during apoptosis by immunocytochemistry.

The fragmentation of chromosomal DNA into oligonucleosomal-sized fragments is one of the hallmarks of apoptosis. Although this can be detected in nerve growth factor (NGF)-deprived sympathetic neurons by running chromosomal DNA on agarose gels *(1)*, TUNEL is a more convenient technique *(2)*. TUNEL is based on the following principle: neurons grown on glass coverslips are fixed and permeabilized and then incubated with terminal deoxynucleotidyl transferase (TdT) and fluorescein-conjugated dUTP. In a tailing reaction, TdT incorporates the labeled dUTP into apoptosis-induced DNA strand breaks *in situ*. After the fragmented DNA has been labeled, a bisbenzimide staining step is performed. This stains chromosomal DNA and enables nuclear morphology to be observed. After mounting on glass slides, the stained neurons are examined on a fluorescence microscope (**Fig. 1**).

An important event in the NGF withdrawal-induced death pathway is the release of cytochrome c from the mitochondria, which precedes DNA fragmentation and chromatin condensation *(3,4)*. Cytochrome c release is necessary for NGF withdrawal-induced apoptosis because cytochrome c is required for caspase activation in the cytosol. Cytochrome c redistribution can be visualized by performing immunocytochemistry with an anticytochrome c antibody *(3)*. In nonapoptotic sympathetic neurons, cytochrome c is localized in the space between the inner and outer mitochondrial membranes. Upon immunostaining, this appears as a bright, punctate cytoplasmic pattern, the nuclear space being essentially clear (**Fig. 2**). After NGF deprivation, cells in which cytochrome c has been released from the mitochondria have a different staining pattern: anti-cytochrome c immunoreactivity is diffuse, faint, and the nuclear space can no longer be easily distinguished (**Fig. 2**). If a tetramethyl rhodamine isothiocyanate (TRITC)-conjugated secondary antibody is used, cytochrome c staining can be combined with TUNEL analysis (fluorescein label) and bisbenzimide staining (blue). Cytochrome c staining is one example of how immunocytochemistry can be used to study the biochemical events that

bisbenzimide TUNEL

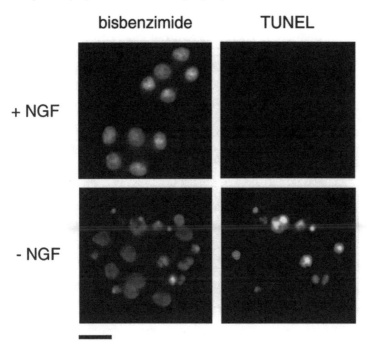

Fig. 1. Apoptotic chromatin condensation and DNA fragmentation in sympathetic neurons visualized by bisbenzimide staining and TUNEL analysis. Sympathetic neurons were cultured on glass coverslips in the presence of NGF for 6 d and then refed with fresh SCG medium containing NGF (+NGF) or lacking NGF (–NGF) as indicated. At 24 h, the cells were fixed, and TUNEL analysis and staining with bisbenzimide (Hoechst) dye were performed as described in **Subheading 3**. In the presence of NGF, the nuclei stain evenly with bisbenzimide dye, and a very low percentage are TUNEL positive. After NGF withdrawal, bisbenzimide staining reveals numerous condensed, distorted, or fragmented nuclei, many of which are TUNEL positive. The bar represents 25 μM.

occur during sympathetic neuron apoptosis. Other events that can be studied by the same type of approach using appropriate antibodies are c-Jun induction, c-Jun phosphorylation, Bax translocation, and caspase 3 activation (5–8).

bisbenzimide   anti-cytochrome c

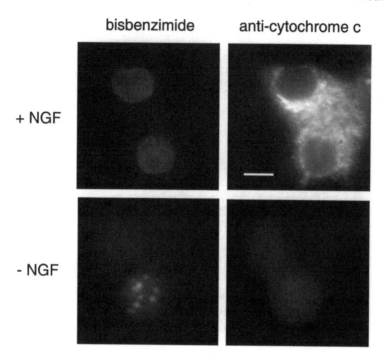

+ NGF

- NGF

Fig. 2. Distribution of cytochrome *c* in normal and apoptotic sympathetic neurons visualized by immunocytochemistry. Sympathetic neurons were cultured on glass coverslips in the presence of NGF for 6 d and then refed with fresh SCG medium containing NGF (+NGF) or lacking NGF (–NGF) as indicated. In the presence of NGF, cytochrome *c* immunoreactivity is excluded from the nuclear space and has a punctate pattern. In the absence of NGF, a fainter, diffuse staining pattern that occurs throughout the whole cell is observed. The bar represents 10 μ*M*.

2. Materials

2.1. TUNEL and Bisbenzimide Staining

1. 4% paraformaldehyde: add 4 g of paraformaldehyde to 100 mL of phosphate-buffered saline (PBS). In a fume hood, stir the solution and carefully heat to 60°C. Add a few drops of 1 *M* NaOH to help dissolve. Cool to room temperature, filter and store as 1-mL aliquots at –20°C.
2. 0.5% Triton X-100 in PBS.

3. *In situ* cell death detection kit, fluorescein (Roche, 1 684 795).
4. Bisbenzimide (Hoechst 33342) trihydrochloride (Calbiochem, 382065) solution: dissolve bisbenzimide at 1 mg/mL in deionized water and store as small aliquots at –20°C.
5. Citifluor glycerol/PBS solution (AF1).
6. Clear nail varnish.

2.2. Cytochrome c Staining

1. 6% paraformaldehyde: add 6 g of paraformaldehyde to 100 mL of PBS (-Ca/Mg) and heat to 50–55°C with stirring to dissolve. Add a few drops of 1 N NaOH to help dissolve. Cool to room temperature and add 0.01% Ca/Mg (90 µL of 1 M $CaCl_2$ and 105 µL of 1 M $MgCl_2$ per 100 mL). Freeze at –20°C in 1-mL aliquots.
2. Blocking/permeabilization solution: 50% goat serum, 0.5% Triton X-100, 0.2% gelatin, 0.5% bovine serum albumin, 0.5× PBS. Add 0.4 g of gelatin (Sigma: swine skin type II, G-2625) to 50 mL of PBS and heat (in a microwave) to dissolve. Allow to cool and add 1 g of bovine serum albumin (Sigma, A-7906), Triton X-100 to 2%, and make up the volume to 100 mL with PBS. Add an equal volume of goat serum and store at 4°C.
3. Mouse anti-cytochrome c monoclonal antibody: clone 6H2.B4 (PharMingen, 65971A). Store at 4°C.
4. Fluorescein-conjugated donkey anti-mouse IgG (H+L) secondary antibody (Jackson ImmunoResearch, 715-095-150): the freeze-dried antibody is reconstituted according to the manufacturer's instructions and stored in small aliquots at –20°C.
5. Bisbenzimide (Hoechst 33342) trihydrochloride (Calbiochem, 382065) solution: dissolve bisbenzimide at 1 mg/mL in deionized water and store as small aliquots at –20°C.
6. Citifluor glycerol/PBS solution (AF1).
7. Clear nail varnish (Boots, UK).

3. Methods

3.1. TUNEL and Bisbenzimide (Hoechst 33342) Staining

For TUNEL and bisbenzimide staining, grow the sympathetic neurons on 13-mm diameter glass coverslips placed in 3.5-cm tissue culture dishes, as described in the previous chapter. 100 μL of each solution is used per coverslip unless stated otherwise.

1. Carefully pick up each coverslip using fine forceps, for example, Dumont no. 5 forceps, and drain off the culture medium by carefully touching the side of the coverslip onto a sheet of clean paper tissue on a flat surface. Be careful not to touch the neurons.

2. Wash each coverslip by dipping it into PBS (fill three wells of a 6-well tissue culture dish with PBS) and then drain away the PBS by touching some clean paper tissue. Wash two more times.

3. Place each coverslip on a level support slightly smaller than the coverslip, which should be firmly fixed in a 9-cm tissue culture dish. The coverslips can be supported on lids cut off microfuge tubes or the adhesive foam spacer blocks that come with the BRL S2 sequencing apparatus can be cut into small squares. Fix the cells by pipetting 100 μL of 4% paraformaldehyde onto each coverslip. Incubate for 30 min at room temperature.

4. Wash in PBS.

5. Permeabilize with 0.5% Triton X-100 in PBS for 5 min at room temperature.

6. Wash in PBS.

7. TUNEL labeling: prepare the TUNEL reaction mixture just before use and store on ice until needed. Mix 50 μL of solution from bottle 1 (10× TdT solution) with 450 μL from bottle 2 (1× label solution). Use 50 μL of TUNEL reaction mixture per coverslip (make sure it spreads evenly). Incubate at 37°C for 60 min in a wet box in the dark.

8. Rinse three times with PBS.

9. Bisbenzimide staining: treat with bisbenzimide dye diluted to 10 μg/mL in deionized water for 5 min at room temperature in the dark.

10. Wash with water twice.

11. For each coverslip, place a drop of Citifluor mounting solution on a glass microscope slide and carefully place the inverted coverslip on it. Gently press the coverslip down onto the microscope slide and wipe away the excess mounting solution that has been squeezed out using a paper tissue. Use clear nail varnish to seal the edges.

12. After the nail varnish has dried, examine the slides on a fluorescence microscope. The chromosomal DNA will have been stained blue by the bisbenzimide dye. Nuclei where DNA fragmentation has occurred will have been stained green from the incorporation of fluo-

rescein-labeled dUTP. The slides should be stored at 4°C in the dark (*see* **Note 2**).

3.2. Cytochrome c Immunocytochemistry

Sympathetic neurons grown on 13-mm diameter glass coverslips placed in 3.5-cm tissue culture dishes are used for cytochrome *c* immunocytochemistry. 100 μL of each solution is used per coverslip unless stated otherwise.

1. Add an equal volume of 6% paraformaldehyde solution to each tissue culture dish and fix for 20 min.
2. Wash with 10 m*M* glycine in PBS.
3. Incubate in blocking/permeabilization solution (50% goat serum, 0.5% Triton X-100, 0.2% gelatin, 0.5% BSA, 0.5× PBS) for 30 min at room temperature.
4. Place the coverslip on a level support as described in **Subheading 3.1., step 3** and add the anti-cytochrome *c* monoclonal antibody diluted 1:100 in blocking/permeabilization solution. Incubate for 2–4 h at room temperature or overnight at 4°C.
5. Wash in PBS.
6. Add the fluorescein-conjugated anti-mouse IgG secondary antibody diluted 1:100 in blocking/permeabilization solution. Incubate for 60 min at room temperature in the dark.
7. Rinse three times with PBS.
8. Bisbenzimide staining: treat with bisbenzimide dye diluted to 10 μg/mL in deionized water for 5 min at room temperature in the dark.
9. Wash with water twice.
10. For each coverslip place a drop of Citifluor mounting solution on a glass microscope slide and carefully place the inverted coverslip on it. Gently press the coverslip down onto the microscope slide and wipe away the excess mounting solution that has been squeezed out using a paper tissue. Use clear nail varnish to seal the edges.
11. After the nail varnish has dried, examine the slides on a fluorescence microscope. The chromosomal DNA will have been stained blue by the bisbenzimide dye. The cytochrome c specific immunostaining will be green because of the binding of the anti-cytochrome *c* primary antibody and fluorescein-conjugated secondary antibody. The slides should be stored at 4°C in the dark.

4. Notes

1. Sympathetic neurons grown on glass coverslips are very easily detached if handled clumsily. It is important to not tear the network of neurites that anchor the neurons to their substrate.
2. Fixation, permeabilization, and blocking conditions could be optimized for individual antigen/antibody interactions by varying the lengths of time for each of these procedures. In general, increasing the time of fixation, permeabilization and blocking may improve results if the conditions described here do not give good results.

Acknowledgments

This work was supported by the Eisai company of Japan and the Wellcome Trust. J.H. is a Wellcome Trust Senior Research Fellow in Basic Biomedical Science.

References

1. Deckwerth, T. L. and Johnson, E. M. (1993) Temporal analysis of events associated with programmed cell death (apoptosis) of sympathetic neurons deprived of nerve growth factor. *J. Cell. Biol.* **123,** 1207–1222.
2. Gavrieli, Y., Sherman, Y., and Ben-Sasson, S. A. (1992) Identification of programmed cell death in situ via specific labeling of nuclear DNA fragmentation. *J. Cell. Biol.* **119,** 493–501.
3. Neame, S. J., Rubin, L. L., and Philpott, K. L. (1998) Blocking cytochrome c activity within intact neurons inhibits apoptosis. *J. Cell. Biol.* **142,** 1583–1593.
4. Deshmukh, M. and Johnson, E. M. (1998) Evidence of a novel event during neuronal death: development of competence-to-die in response to cytoplasmic cytochrome c. *Neuron* **11,** 695–705.
5. Ham, J., Babij, C., Whitfield, J., Pfarr, C. M., Lallemand, D., Yaniv, M., and Rubin, L. L. (1995) A c-Jun dominant negative mutant protects sympathetic neurons against programmed cell death. *Neuron* **14,** 927–939.
6. Eilers, A., Whitfield, J., Babij, C., Rubin, L. L., and Ham, J. (1998) Role of the Jun kinase pathway in the regulation of c-Jun expression and apoptosis in sympathetic neurons. *J. Neurosci.* **18,** 1713–1724.

7. Putcha, V., Deshmukh, M., and Johnson, E. M. (1999) Bax translocation is a critical event in neuronal apoptosis: regulation by neuroprotectants, Bcl-2 and caspases. *J. Neurosci.* **19,** 7476–7485.
8. Finn, J. T., Weil, M., Archer, F., Siman, R., Srinivasan, A., and Raff, M. C. (2000) Evidence that Wallerian degeneration and localised axon degeneration induced by local neurotrophin deprivation do not involve caspases. *J. Neurosci.* **10,** 1333–1341.

13

Measuring Programmed Cell Death in Plants

Ludmila Rizhsky, Vladimir Shulaev, and Ron Mittler

Summary

Methods for the detection of programmed cell death (PCD) in plants are reviewed with references for different biochemical, microscopic, and molecular assays. A detailed description of three different methods for the detection of biotic or abiotic PCD in plant tissues is included. The reader is encouraged to use all three methods in parallel to obtain a reliable measure of PCD. Critical considerations are highlighted.

Key Words: Programmed cell death; plant; hypersensitive response; nuclease; DNA degradation; ion lekeage.

1. Introduction

Several methods are currently being used to measure programmed cell death (PCD) in plants. These can be divided into biochemical and microscopic. Microscopic methods will not be described in this chapter. They include staining of single plant cells in culture with a dye indicating their viability *(1)* and viability staining of whole plant tissues *(2)*, which enables the detection of dead cells within leaves. PCD in plants can also be followed microscopically by the use of the classical terminal deoxynucleotidyl transferase-mediated dUTP nick end labeling staining (in combination with DAPI staining), which detects the presence of degraded DNA within nuclei *(3)*.

From: *Methods in Molecular Biology, vol. 282: Apoptosis Methods and Protocols*
Edited by: H. J. M. Brady © Humana Press Inc., Totowa, NJ

However, some modifications may be required to convert the commercial kits available for terminal deoxynucleotidyl transferase-mediated dUTP nick end-labeling staining in animals to the staining of PCD in plants *(4)*. It is also useful to include a positive control, such as differentiating tracheary elements *(5)*. Microscopic observation of PCD in plants is often easier compared with the observation of apoptosis in animals. The reason for this is that the corpse of dead plant cells or at least the plant cell walls are typically not eliminated after death. Nonetheless, a large variability in morphological changes was observed between different types of PCD in plants, and establishing a common set of morphological characteristics, similar to apoptosis in animals, is not possible at present *(6)*.

Although complete sequencing of the *Arabidopsis thaliana* genome did not reveal apparent homologs of apoptotic proteins, including caspases, in plants, recent data indicate that several plant proteins have regions with similarity to mammalian proteins involved in cell death *(7,8)*. Constitutive caspase-like machinery has been postulated to be responsible for the execution of certain types of PCD in plants *(9)*. Additionally, an ancient family of caspase-like proteins, namely metacaspases, was identified in the *Arabidopsis* genome using caspase-like domains of the *Dictyostelium* sequence as a query for BLAST search *(10)*. Several cysteine proteases may also be involved in pathogen-mediated PCD in plants. Microarray analysis of global gene expression during various types of PCD may help to identify a set of genes that are common to all types of PCD or that are differentially expressed during specific PCD process in plants. Custom microarray analysis of gene expression during PCD induced in *Arabidopsis* cell culture by heat treatment or during senescence revealed subsets of genes specific to a particular treatment, as well as a set of genes induced by both treatments *(11)*. Some of these transcripts, induced during different types of PCD (e.g., cysteine proteinases, several transcription factors, and hypersensitive response-related genes), may serve as markers for the core PCD program in plants. Identification of cell death-related proteins, including specific proteases, and exact mechanism of their regulation would facilitate the development of new assays for early

PCD detection in plant cells and tissues. However, at present these molecular markers are only reliable within a defined set of PCD processes. Biochemical methods for the detection of PCD in plants include the isolation of DNA from tissues and the detection of DNA fragmentation via conventional (DNA ladder) or field inversion agarose gel electrophoresis (FIGE, 50- or 300-kb fragments), the isolation of protein extracts and the detection of nuclease or protease activity in total or nuclei extracts by activity gels, and measuring of ion leakage from leaf discs *(4,12–14)*. The detection of ion leakage from cells is a relatively good measure of cell death in plants because plant cells do not form apoptotic bodies and the content of the dead cell is typically released into the intercellular fluid. In contrast, some controversy has arisen with respect to the use of DNA ladders to measure PCD in plant cells *(15)*. In this chapter we describe a combined biochemical approach to measuring PCD in plants that uses the detection of large DNA fragments by FIGE, the detection of nuclease activation by nuclease activity gels, and the detection of ion leakage from leaf discs.

2. Materials

1. Protein extraction buffer: 100 mM MOPS, pH 6.8; 5 mM ascorbate; 2 mM reduced glutathione; 1 mM CaCl$_2$; 0.5 mM MgCl$_2$; 0.5 mM phenylmethylsulfonyl fluoride; 1.5% sodium dodecyl sulfate (SDS); and 150 mM Tris-HCl (pH 6.8).
2. Gel wash solution 1: 25% isopropanol; 10 mM MOPS, pH 6.8; 0.1 mM CaCl$_2$; 0.1 mM MgCl$_2$.
3. Gel wash solution 2: 10 mM MOPS, pH 6.8, 1 mM CaCl$_2$, 0.5 mM MgCl$_2$.
4. Gel incubation buffer: 10 mM MOPS, pH 6.8, 1 mM CaCl$_2$, 0.1 mM MgCl$_2$.
5. ET buffer: 10 mM Tris-HCl, pH 8.0, 50 mM ethylenediamine tetraacetic acid (EDTA).
6. FIGE plug incubation buffer: 10 mM Tris-HCl, pH 8.0, 0.5 M EDTA, 1% sarcosyl, and 1 mg/mL proteinase K (Promega).
7. TBE buffer: 45 mM Tris-borate, 1 mM EDTA, pH 8.0.
8. 2× SSC: 30 mM Na-citrate, pH 7.0, 300 mM NaCl.
9. DNA hybridization buffer: 0.25 M Na$_2$HPO$_4$, pH 7.4, 7% SDS, 1% casein, 1 mM EDTA (heat to dissolve).

3. Methods

To measure PCD, plants or leaves should be treated with a specific PCD inducer, such as a chemical *(4,14)* or a pathogen *(12, 13,16)*, as well as mock treated, and sampled for analysis. Leaf samples should be divided into fresh plant material in the form of leaf discs to be used for ion leakage measurements and frozen plant material (flash frozen in liquid nitrogen) to be used for protein or DNA isolation. To collect identical amounts of plant material for these analyses, for each sample a set number of leaf discs can be pooled, weighed, and frozen.

3.1. Measurement of Ion Leakage From Leaf Discs

1. Prepare leaf discs (9-mm diameter) by placing leaves abaxial side down on a Styrofoam surface and gently punching the leaf using a commercial cork puncher between the veins (leaf discs should not contain major leaf veins since these can interfere with the measurement; *see* **Note 1**).
2. For each measurement, float five leaf discs abaxial side up on 5 mL of double-distilled water in a small Petri dish for 3 h at room temperature. Before the 3 h incubation time in 5 mL of water, the leaf discs should be floated for 5 min in water (50 mL) in a separate dish to avoid measuring ion leakage related to the immediate injury inflicted on the plant tissue during the preparation of leaf discs (*see* **Notes 2** and **3**).
3. After the 3 h incubation period, transfer the bathing solution to a tube and measure its conductivity with a conductivity meter (i.e., model 604; VWR Scientific). Refer to this value as value A.
4. Introduce the leaf discs and the bathing solution into sealed tubes and incubate at 95°C for 25 min (this treatment will cause the complete leakage of ions from cells).
5. Cool to room temperature.
6. Measure again the conductivity of the bathing solution, and refer to this value as value B.
7. Express ion leakage as % leakage, that is (value A/ value B) × 100.
8. The measurement of ion leakage for each time point in a given experiment should be performed in triplicates (i.e., three Petri dishes with five leaf discs in each). **Figure 1A** shows an example of mea-

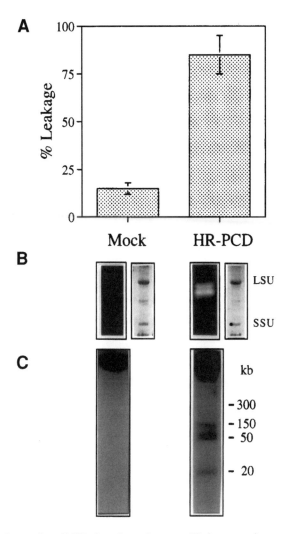

Fig. 1. Measuring PCD in plant leaves. Tobacco plants were inoculated with tobacco mosaic virus strain U1 in 5 mM potassium phosphate buffer, pH 7 (HR-PCD), or mock infected with 5 mM phosphate buffer (Mock) by gently rubbing the leaves with carborundum and kept at 30°C for 5 d. PCD was induced by shifting of plants to 22°C *(5)*. Leaves of Mock and tobacco mosaic virus-infected plants were collected 24 h after the temperature shift and assayed for PCD by measuring ion leakage (**A**), nuclease activity (**B**), and DNA fragmentation (**C**). HR, hypersensitive response; LSU and SSU, Rubisco large and small subunits respectively; PCD, programmed cell death.

suring ion leakage in a control (mock) and a pathogen-treated (HR-PCD) plant (24 h after mock or pathogen infection). For each treatment, three replicates were performed and the data is presented as the mean and standard deviation of % leakage.

3.2. Measuring Nuclease Activity

3.2.1. Protein Extraction

1. To extract soluble leaf proteins, grind the frozen leaf discs in liquid nitrogen to a fine powder with a mortar and a pestle and dissolve the powder in the Protein extraction buffer (*see* **Note 4**).
2. Heat the protein extracts to 65°C for 10 min and centrifuged for 5 min at 10,000*g*.
3. To estimate protein content, recover the protein extracts from the supernatant and subject them to SDS-polyacrylamide gel electrophoresis (PAGE) followed by staining with Coomassie blue. Alternatively, use a protein determination assay that is not sensitive to the presence of SDS in protein extracts.

3.2.2. Detection of Nuclease Activity With Activity Gels

1. Subject protein extracts to SDS-PAGE in gels that were cast with a single- or double-strand DNA substrate (50 µg/mL of heat-denatured or native salmon sperm DNA) and 50 µg/mL of bovine fibrinogen. Use a prestained molecular weight marker to estimate the apparent size of the electrophoretically separated nucleases.
2. After electrophoresis, wash the gels twice for a total of 30 min with the gel wash solution 1 to remove SDS from the gels, and twice for a total of 60 min with the gel wash solution 2 (*see* **Note 5**).
3. Incubate the gels overnight at 37°C or 45°C in the gel incubation buffer, with gentle agitation (*see* **Note 6**).
4. Stain the gels with 0.02% toluidine blue (Sigma) for 10 min and wash extensively with 10 m*M* MOPS, pH 6.8. Nuclease activity is detected as an achromatic region (band) that does not stain with toluidine blue because of the absence of the DNA substrate.
5. To demonstrate equal loading of protein samples on the nuclease activity gels, duplicate gels that do not contain DNA substrate and fibrinogen should be stained with Coomassie Blue. **Figure 1B** shows an example of measuring nuclease activity in a control (mock) and a pathogen-treated (HR-PCD) plant (24 h after mock or pathogen

infection). For each treatment, three replicates were performed and the protein extracts were pooled. The data are presented as a nuclease activity gel of the pooled extracts (left panel for each treatment) and a duplicate Coomassie-stained regular SDS gel of the pooled extracts (right panel for each treatment) to show equal loading of proteins on the nuclease activity gel.

3.3. Detecting DNA Fragmentation by FIGE

3.3.1. Tissue Preparation and Running FIGE Gels

1. Ground frozen leaf discs (0.2 g) to a fine powder in liquid nitrogen.
2. Dissolve the powder in 1 mL of ET buffer preheated to 50°C by gently inverting the tube.
3. Mix with 1 mL of 1.4% (w/v) low melting point agarose (BRL) prepared in ET buffer and cooled to 50°C by gently inverting.
4. Pour agarose plugs in an FIGE plug mold placed on ice and let harden (15–20 min).
5. Incubate plugs for 3 h in FIGE plug incubation buffer at 52–54°C with gentle agitation (*see* **Note 7**).
6. Wash plugs with ET buffer (3 × 10 min) and store at 4°C in ET buffer (up to 1 wk).
7. Place plugs in gel, seal with 0.7% (w/v) low melting point agarose (BRL) prepared in ET buffer and cooled to 50°C, and perform an FIGE run. Two examples are given below:
   a. Use a Bio-Rad CHEF Mapper™ apparatus with a 1% agarose (BRL) gel and 0.5× TBE buffer (maintained at 10°C). To determine the size of DNA fragments, use a 50-kb DNA ladder (Promega). Perform electrophoretic separation with a 6 V/cm voltage gradient at an angle of 120° and a linear ramping factor with alternating pulse time of 5.3 to 20.5 s over the course of 18 h.
   b. Use a Hofer Switchback™ Pulse Controller with a 1% agarose gels and 0.5× TBE buffer (maintained at 15°C). Perform electrophoretic separation with a 6.9 V/cm using a reverse mode with an F/R ratio of 3.0:1 and a multiple run mode (run 1: 10 min run-in followed by 8 h at a pulse time of 1 to 20 s; run 2: 8 h at a pulse time of 0.8 to 1.5 s).
9. Stain the gels with ethidium bromide and photograph. The loading of DNA onto the FIGE gel is based upon equal amounts of tissue fresh weight. On average, each plug should contain about 0.025 g of tissue.

3.3.2. Transfer of DNA From FIGE Gels and Hybridization

1. Depurinate DNA in FIGE gels by incubating gels in 0.25 N HCl for 10–15 min.
2. Wash gels with double distilled water (3 × 15 min).
3. Incubate gels in 0.4 N NaOH, 1.5 M NaCl for 20 min with gentle agitation.
4. Transfer DNA to a nylon membrane (Zeta-Probe GT, Bio-Rad) by the capillary transfer method using 0.4 N NaOH, 1.5 M NaCl for 48 h.
5. Wash the membrane briefly in 2× SSC and bake in a vacuum oven at 80°C for 30 min.
6. Hybridize the membrane with a mixture of probes corresponding to different nuclear genes to detect fragmentation of nuclear DNA, and/or a mixture of probes corresponding to different chloroplast genes to detect fragmentation of chloroplast DNA or a mixture of probes corresponding to different mitochondrial genes to detect fragmentation of mitochondria DNA. Use the DNA hybridization buffer and label probes with a random-primed labeling kit using ^{32}P-dATP (*see* **Note 8**).
7. Wash the membrane (4 × 15 min with 1× SSC, 0.1% SDS at room temperature), expose to an X-ray film, and develop. **Figure 1C** shows an example of measuring DNA fragmentation by an FIGE gel in a control (mock) and a pathogen-treated (HR-PCD) plant (24 h after mock or pathogen infection). The data are presented as an X-ray film showing the fragmentation of nuclear (50 kb), chloroplast (150 kb), and mitochondria (20 and 50 kb) DNA during HR-PCD (right panel) and the absence of DNA fragmentation in control (Mock-treated plants; left panel).

4. Notes

1. When preparing leaf discs, it is imperative to avoid major vein tissue because this tissue may contain plant cells of the vascular system, which undergo developmentally regulated PCD independent of the experimental procedure.
2. Leaf discs should be kept during incubation under the same light intensity as the plants from which they were taken to avoid interference with the rate of PCD (most pathogen-induced PCD processes can be inhibited in the dark).

3. When dealing with small plants, such as *Arabidopsis*, the entire leaf can be floated on water to determine ion leakage (% leakage takes into account the total possible leakage from the tissue hence, enables the comparison of ion leakage from two different leaves with slightly different sizes).

4. The detection of nuclease activity in total extracts can be complicated because of the large number of nucleases that are activated during PCD and/or related processes such as wounding *(13)*. It is therefore suggested that at least in the initial stages of setting up the experimental system nuclease activity will be compared between total leaf extracts and protein extracts obtained from isolated nuclei *(12)*. This should help in determining whether the major nucleases detected in the total crude extracts represent the nucleases, which are responsible to the degradation of DNA within the nuclei. It is important to remember however, that in plants the degradation of nuclear DNA may occur only during late stages of PCD when the integrity of the vacuole and/or plasma membrane has been compromised *(6,16)*.

5. The use of nuclease activity gels for the detection of nuclease activity allows studying the effect of different ions or inhibitors on the activity of nucleases. These may be included in the gel wash solutions and the gel incubation buffer *(12)*.

6. Shorter or longer incubation times, as well as different incubation temperatures, may be needed to detect different nucleases with the nuclease gel assay. It is therefore recommended to vary these conditions during the initial setup of the experimental system for a more comprehensive characterization.

7. For different plants different incubation times of FIGE plugs in the FIGE plug incubation buffer may be required (up to overnight) because of differences in cell wall and/or other cellular structures.

8. The transfer of DNA from FIGE gels and the probing of membranes with different probes considerably enhance the sensitivity of this method. In addition they allow the study of DNA degradation in different organelles during PCD. For example, it was found that chloroplast DNA degradation is the first DNA degradation process that accompanies TMV-induced PCD in tobacco *(16)*.

Acknowledgments

We wish to thank Dr. Eric Lam for sharing of unpublished data. This work was supported by funding provided by The Israel Ministry of Agriculture, The Yigal Alon Fellowship, and The Hebrew University Intramural Research Fund Basic Project Awards.

References

1. Levine, A., Tenhaken, R., Dixon, R., and Lamb, C. (1994) H_2O_2 from the oxidative burst orchestrates the plant hypersensitive disease resistance response. *Cell* **19**, 583–593.
2. Weymann, K., Hunt, M., Uknes, S., Neuenschwander, U., Lawton, K., Steiner, H. Y., and Ryals, J. (1995) Suppression and restoration of lesion formation in Arabidopsis *lsd* mutants. *Plant Cell* **7**, 2013–2022.
3. Gavrieli, Y., Sherman, Y., and Ben-Sasson, S. A. (1992) Identification of programmed cell death in situ via specific labeling of nuclear DNA fragmentation. *J. Cell Biol.* **119**, 493–501.
4. Wang, H., Li, J., Bostock, R. M., and Gilchrist, D. G. (1996) Apoptosis: a functional paradigm for programmed plant cell death induced by a host-selective phytotoxin and invoked during development. *Plant Cell* **8**, 375–391.
5. Mittler, R., Shulaev, V., and Lam, E. (1995) Coordinated activation of programmed cell death and defense mechanisms in transgenic tobacco plants expressing a bacterial proton pump. *Plant Cell* **7**, 29–42.
6. Mittler, R. (1998) Cell death in plants, in Zakeri, Z., Tilly, J., Lockshin R. A. (eds.), *When Cells Die, A Comprehensive Evaluation of Apoptosis and Programmed Cell Death*, John Wiley & Sons, New York, pp. 147–174.
7. Aravind, L., Dixit, V. M., and Koonin, E. V. (1999) The domains of death: evolution of the apoptosis machinery. *Trends Biochem. Sci.* **14**, 47–53.
8. Lam, E., Kato, N., and Lawton, M. (2001) Programmed cell death, mitochondria and the plant hypersensitive response. *Nature* **111**, 848–853
9. Elbaz, M, Avni, A, and Weil, M. (2002) Constitutive caspase-like machinery executes programmed cell death in plant cells. *Cell Death Differ.* **9**, 726–733.
10. Uren, A. G., O'Rourke, K., Aravind, L., Pisabarro, M. T., Seshagiri, S., Koonin, E. V., et al. (2000) Identification of paracaspases and metacaspases: two ancient families of caspase-like proteins, one of which plays a key role in MALT lymphoma. *Mol. Cell* **6**, 961–967.

11. Swidzinski, J. A., Sweetlove, L. J., and Leaver C. J. (2002) A custom microarray analysis of gene expression during programmed cell death in *Arabidopsis thaliana*. *Plant J.* **10,** 431–46.

12. Mittler, R. and Lam, E. (1995) Identification, characterization, and purification of a tobacco endonuclease activities induced upon hypersensitive response cell death. *Plant Cell* **7,** 1951–1962.

13. Mittler, R. and Lam, E. (1997) Characterization of nuclease activities and DNA fragmentation induced upon hypersensitive response cell death and mechanical stress. *Plant Mol. Biol.* **14,** 209–221.

14. Ryerson, D. E. and Heath, M. C. (1996) Cleavage of nuclear DNA into oligonucleosomal fragments during cell death induced by fungal infection or by abiotic treatments. *Plant Cell* **8,** 393–402.

15. Fath, A. Bethke, P. C., and Jones, R. L. (1999) Barley aleurone cell death is not apoptotic: characterization of nuclease activities and DNA degradation. *Plant J.* **10,** 305–315.

16. Mittler, R., Simon, L., and Lam, E. (1997) Pathogen-induced programmed cell death in tobacco. *J. Cell Sci.* **110,** 1333–1344.

14

Detection of Apoptosis in *Drosophila*

Kimberly McCall and Jeanne S. Peterson

Summary

Drosophila has unique genetic and cell biological advantages as a model system for the study of apoptosis. Many cell death genes are evolutionarily conserved between flies and mammals. Cell death can be induced by environmental stimuli and normally occurs during diverse developmental processes in *Drosophila*. Here, we review several approaches for detecting cell death in *Drosophila*. We provide detailed protocols for labeling apoptotic cells in the embryo and ovary using terminal deoxynucleotidyl transferase-mediated dUTP nick end labeling and acridine orange. Additionally, we describe methods for ectopically expressing cell death genes in the eye and the use of transgenic flies for the detection of genetic interactions among cell death genes.

Key Words: *Drosophila*; apoptosis; TUNEL; acridine orange; cell death.

1. Introduction

Drosophila has unique genetic and cell biological advantages as a model system for the study of apoptosis *(1,2)*. Cell death occurs in diverse developmental processes in *Drosophila*, including the formation of the embryonic nervous system, the destruction of larval tissues during metamorphosis, the morphogenesis of the eye, and the generation of eggs in the ovary *(3–6)*. Additionally, cells die ectopically in response to environmental stimuli, such as X-rays *(4)*.

From: *Methods in Molecular Biology, vol. 282: Apoptosis Methods and Protocols*
Edited by: H. J. M. Brady © Humana Press Inc., Totowa, NJ

Genetic studies in *Drosophila* have uncovered three novel cell death activators, *reaper*, *hid*, and *grim* *(7–9)*. Flies homozygous for the *H99* deletion, which removes these three genes, die during embryogenesis with a complete block in apoptosis *(7)*. Although the *H99* genes are essential for embryonic apoptosis, they are not required for the cell death of ovarian nurse cells, suggesting that there exists at least one other cell death pathway in flies *(10)*. Overexpression of *reaper*, *hid*, or *grim* leads to widespread apoptosis, which can be blocked by the overexpression of the caspase inhibitors p35 or inhibitor of apoptosis proteins (IAPs; **refs. 8,9,11,12**). Reaper, Hid, and Grim are thought to function by generally inhibiting protein translation and/or triggering the auto-ubiquitination of IAPs, reducing the level of IAPs such that caspases are not inhibited (reviewed in **ref. 13**). Consistent with this model, homozygous loss-of-function *thread* (*DIAP1*) mutants die early in embryogenesis with extensive apoptosis (reviewed in **ref. 1**).

The genome sequence of *Drosophila* has been completed, revealing fly homologs for most mammalian cell death genes. Genes encoding seven caspases, two Bcl-2 family members, four BIR domain proteins, and one each of Apaf-1, FADD, CAD, p53, tumor necrosis factor (TNF), and TNFR have been identified in flies *(1,2,14–16)*. Loss-of-function phenotypes have been generated for several of these genes by transposon insertion, RNAi, expression of dominant negative alleles, or gene targeting (reviewed in **refs. 1,14,16–18**).

One of the strengths of the *Drosophila* system is the ability to use cell biological techniques on a variety of developmental stages. Expression of cell death genes can be visualized by whole-mount RNA *in situ* hybridization or immunocytochemistry. Commercial antibodies are rarely available for *Drosophila* proteins; however, two active caspase-3 antibodies have been shown to cross-react with *Drosophila* caspases *(19,20)*. Antibodies have also been described for some *Drosophila* cell death proteins, such as Hid, Thread, cytochrome *c*, Drice, and Dronc *(21–24)*. Additionally, the detection of macrophages is correlated with the amount of cell death, and macrophages can be visualized by antibodies such as those against Peroxidasin and Croquemort *(4,25–27)*. Besides being able to study cell death in the intact fly, several *Drosophila* cell lines exist that can be subjected to the same analysis as mammalian cell lines *(28)*.

In this chapter, we describe three methods typically used for the detection of apoptosis in *Drosophila*. The terminal deoxynucleotidyl transferase-mediated dUTP nick end labeling (TUNEL) and acridine orange (AO) methods are used to detect dead or dying cells in a variety of tissues. We focus on methods for the embryo and ovary, but these techniques can be used on other tissues as well. The third method is the detection of genetic interactions by expressing cell death genes in the *Drosophila* eye.

2. Materials

The methods described below review general fly handling as well as specific apoptosis techniques. However, workers new to *Drosophila* should consult with a *Drosophila* laboratory before beginning to work with flies.

2.1. Fly Handling

1. Instant fly food may be purchased from Fisher Scientific or Carolina Biological (Burlington, NC). Fly food should be poured into vials or bottles (available from Fisher or VWR).
2. Dissecting microscope with a 6.5–50× magnification range.
3. Diffuser pad for CO_2 anesthesia (Fisher) or shaved ice and dry vials for cold anesthesia.
4. Baskets with 80-μ nitex mesh (Sefar America, Kansas City, MO) for dechorionating embryos. Baskets can be made by attaching a small piece of mesh to one end of a 1-inch long tube.
5. Fine forceps, tungsten needles, and glass plates or depression slides for dissection.
6. Apple juice/agar plates: Mix 90 g of Difco agar with 3 L of water, autoclave for 50 min, and cool in a 60°C water bath. Mix 1 L of apple juice with 100 g of sugar and heat to 60°C to dissolve. Combine agar/water and juice/sugar mixtures, stir, add 60 mL of a 10% solution of *p*-hydroxy benzoic acid methyl ester (Sigma) in ethanol, and pour the plates. The tops of 35 × 10-mm plates (Falcon) will fit fly food bottles from Fisher.

7. Egg laying chambers. These may be made by cutting a hole in the side of a dry fly food bottle and stuffing it with a cotton ball. The apple juice/agar plate will fit on the mouth of the bottle.

8. Yeast paste: Mix granular yeast (Sci-Mart, Inc., St. Louis, MO) and water into a smooth paste.

9. *Drosophila* Ringers (DR): 130 mM NaCl, 4.7 mM KCl, 1.9 mM CaCl$_2$, and 10 mM HEPES, pH 6.9, made 10× and stored frozen.

2.2. TUNEL Materials

1. Phosphate-buffered saline (PBS): 130 mM NaCl, 7 mM Na$_2$HPO$_4$, 3 mM NaH$_2$PO$_4$.

2. PBS with 0.1% Tween-20 (PBT).

3. Heptane.

4. TritonX-100.

5. Proteinase K (Fisher) stock solution, 20 mg/mL in dH$_2$O, stored frozen in 10-µL aliquots.

6. Bovine serum albumin Fraction V (Fisher).

7. Normal goat serum (Gibco BRL, Grand Island, NY).

8. pH 9 buffer: 0.1 M Tris-HCl, pH 9.5, 0.1 M NaCl, 50 mM MgCl$_2$, 0.1% Tween-20.

9. Fixative: 4% paraformaldehyde (Sigma) in PBS. Heat to dissolve.

10. 70% glycerol in PBS.

11. Methanol.

12. Household bleach, for example, Clorox.

13. ApopTag reagents (Serologicals Corp., Norcross, GA): Equilibration buffer, reaction buffer containing nucleotides labeled with digoxigenin (RXB), terminal deoxyribonucleotidyl transferase (TdT), and stop-wash buffer (SWB).

14. Roche reagents (Indianapolis, IN): antiDigoxigenin antibody complexed to alkaline phosphatase, nitroblue tetrazolium salt, and 5-bromo-4-chloro-3-indolyl phosphate toluidinium salt (X-Phos).

2.3. AO Materials

1. Stock solution AO (Sigma A 6014) dissolved in dH$_2$O at 1 mg/mL, stored in dark at 4°C.

2. 0.1 M phosphate buffer, pH 7.0.

3. Heptane.

4. Halocarbon oil (series 700, Halocarbon Products Corporation, River Edge, NJ).
5. Fluorescence microscope equipped with fluorescein, rhodamine and ultraviolet (UV) filters, brightfield/differential interference contrast, and a camera.

2.4. Genetic Interactions Materials

1. pGMR vector *(11)* or other eye-specific vector.
2. pπ25.7wcΔ2-3 plasmid (**ref. 29**, available from many *Drosophila* laboratories).
3. Injection buffer: 5 mM KCl, 0.1 mM NaH$_2$PO$_4$, pH 6.8.
4. Fly microinjection facility.
5. Fly strains. *GMRrpr, GMRhid, GMRp35, GMRdiap1,* and *GMRdiap2* are all available from the *Drosophila* stock center at Bloomington, IN (*see* http://fly.bio.indiana.edu/).
6. A camera attachment for the dissecting microscope is necessary for making a photographic record of eye phenotypes.

3. Methods

3.1. Sample Preparation

For well-developed ovaries and/or good embryo production, start with equal numbers of 3- to 7-d-old male and female flies kept together in uncrowded conditions, transferring them to new food vials supplemented with wet yeast paste once or twice daily for 4 d or more before collecting samples.

3.1.1. Ovary Dissection

1. Anesthetize flies under CO_2 or on ice.
2. Dissect females in depression plates or slides in a drop of DR. Grasp fly between thorax and abdomen with forceps, and pull at the terminal part of the abdomen with another pair of forceps to release the ovaries and other organs from the cavity.
3. Tease ovaries away from debris and separate ovarioles from each other with tungsten needles.

4. Transfer tissue in DR to microcentrifuge tubes (*see* **Notes 1** and **2**). Hold tissues on ice until all samples have been collected. Proceed to **Subheading 3.2.1.** for TUNEL or **Subheading 3.3.** for AO staining.

3.1.2. Embryo Collection

1. Apply a dab of fresh yeast paste to an apple juice/agar plate. Fix the apple juice/agar plate against the mouth of the egg-laying chamber. Transfer flies to the chamber and allow flies to lay eggs for the desired time.
2. Use water and a fine brush to dislodge the embryos from the surface of the plate, and collect them with a large (1000 µL) pipet tip (*see* **Note 2**).
3. Transfer the embryos to baskets and remove the water. Dechorionate embryos in baskets using 50% bleach for 2–5 min and wash several times with water. Proceed to **Subheading 3.2.2.** for TUNEL or **3.3.** for AO staining.

3.2. TUNEL Staining

For TUNEL staining, ovaries or embryos are treated the same as in *in situ* hybridization and antibody staining. The protocol below was derived from the description of ovarian tissue preparation by Verheyen and Cooley *(30)*, from descriptions of embryo staining in protocols 54 and 95 in Ashburner *(29)*, and from the description of TUNEL staining by White et al. *(7,31)*.

3.2.1. Ovary Fixation

1. Mix heptane and fix 5:1, remove DR from ovaries (*see* **Note 3**), add heptane/fix, and agitate gently for 30 min at room temperature.
2. Remove heptane/fix and wash twice with excess PBT, taking care to remove all heptane droplets. Proceed to **Subheading 3.2.3.**

3.2.2. Embryo Fixation

1. Mix heptane and fix 1:1.
2. Transfer embryos to the fixing solution (*see* **Note 4**) and shake for 20 min at room temperature.

3. Remove fix (bottom layer) first and then remove heptane. Add fresh heptane and shake.

4. Add a double quantity of methanol and shake hard (vortex) for 2 min to remove the vitelline membrane.

5. Discard embryos at interface, remove heptane and then methanol, and wash twice with methanol.

6. Rehydrate through a series of 75%, 50%, and 25% methanol in PBT. Proceed to **Subheading 3.2.3.**

3.2.3. General TUNEL Staining Protocol

1. Treat fixed tissue with Proteinase K, 10 µg/mL in PBT (5 min for ovaries, 3 min for embryos) and wash twice with PBT.

2. Post fix for 20 min in a solution of 4% paraformaldehyde in PBS, then wash five times, 5 min each, in PBT.

3. Equilibrate for 1 h at room temperature in equilibration buffer.

4. Incubate overnight at 37°C in a reaction mix consisting of RXB and TdT in a 2:1 ratio, with 0.3% Triton-X 100.

5. Preabsorb antiDig-AP, diluted 1:2000 in PBT, with fixed tissue for 2 h at room temperature or at 4°C overnight.

6. Remove RXB and TdT from tissue and incubate in SWB diluted to 1:34 in water at 37°C for 3–4 h, changing the solution every 20–30 min. Remove SWB and wash three times, 5 min each, in PBT.

7. Block in a solution of 2 mg/mL BSA, 5% normal goat serum in PBT for 1 h at room temperature.

8. Incubate tissue in preabsorbed antibody for 2 h at room temperature or overnight at 4°C.

9. Wash four times for 20 min each in PBT and wash twice, 20 min each, in pH 9.0 buffer.

10. Add 3.5 µL of nitroblue tetrazolium salt and 4.5 µL of 5-bromo-4-chloro-3-indolyl phosphate toluidinium salt to 1 mL of pH 9 buffer and incubate tissues, watching carefully for the color reaction.

11. Stop the reaction with a PBT wash and mount in 70% glycerol.

12. View under brightfield or differential interference contrast microscopy.

3.3. AO Staining

AO is a vital dye that differentially stains living and dying cells. The use of other vital dyes is described elsewhere *(4)*. The advantage of AO is that it is performed quickly on live tissue. However, this also means that the tissue must be examined and photographed immediately after staining. These protocols are derived from protocols described in **refs. 4** and **10**.

3.3.1. Embryo Protocol

1. Dilute AO stock solution to 5 µg/mL in 0.1 *M* phosphate buffer.
2. Collect embryos in mesh baskets as described and wash only in water (*see* **Note 5**). Using a fine-tipped paintbrush, transfer embryos from the mesh to tubes containing an equal volume of heptane and the 5 µg/mL AO solution. Microcentrifuge tubes or glass tubes with tight-fitting caps may be used.
3. Shake tubes vigorously by hand for 3–5 min. Shaking by hand improves the permeability of the embryos (*see* **Note 6**).
4. Pipet off the liquid and replace with heptane.
5. Pipet the embryos in heptane onto glass slides. Try to keep the embryos separated and soak up the heptane using a Kim wipe twisted into a point (*see* **Note 7**). Quickly cover the embryos with halocarbon oil and a coverslip.
6. View the slide immediately under epifluorescence. AO staining is visible under both rhodamine and fluorescein filters. The rhodamine filter often looks better because the fluorescein filter shows more background and smearing from residual heptane.

3.3.2. Ovary Protocol

1. Dilute AO stock solution to 10 µg/mL in phosphate buffer.
2. Transfer dissected ovaries to an Eppendorf tube containing 15 µL of heptane and 15 µL of 10 µg/mL AO solution.
3. Flick tube gently to mix and allow to rotate for 5 min.
4. Transfer ovaries to slides and spread out the ovary tissue into individual egg chambers if possible. Pipet off the AO/heptane mixture or use a Kim wipe twisted into a point (*see* **Note 7**). Cover with halocarbon oil and a coverslip.
5. View the slide immediately under epifluorescence, using the fluorescein, rhodamine or UV filter. Under UV, the apoptotic nuclei stain yellow or red (**10**).

3.4. Detection of Genetic Interactions

The *Drosophila* eye is a commonly studied tissue for many cell biological processes. It has several advantages: it has a repeating structure made up of only a few cell types, the development of this tissue is highly organized, and the eye is a nonessential tissue. Cell death normally occurs in the developing eye disc and can easily be induced in other cells. Several eye-specific expression vectors exist and we will focus on one of these, the pGMR vector *(11)*, which is expressed in all cells in the differentiated portion of the eye disc. Many fly strains have been generated that ectopically express cell death genes in the eye (a partial list is in **Table 1**).

3.4.1. Generation of Transgenic Flies

1. Subclone a gene of interest into the pGMR vector using standard molecular biology techniques.
2. Purify the final construct using CsCl banding or the Qiagen Endofree Plasmid Kit (Qiagen Inc., Valencia, CA).
3. Combine 10 µg of the final pGMR construct with 4 µg of the pπ25.7wcΔ2-3 plasmid (purified with CsCl or Endofree kit), and ethanol precipitate. Resuspend in 20 µL of injection buffer.
4. Inject into the posterior of preblastoderm embryos that are *white*⁻ (this requires a *Drosophila* microinjection facility). Further details on *Drosophila* transformation can be found in Ashburner *(29)*.
5. Transfer larvae that survive the injection procedure to a vial of fly food.
6. Collect the adult flies as they emerge and mate them individually to *white*⁻ flies. The pGMR vector carries the *white*⁺ gene that will mark the transgenic flies. However, the injected flies will only carry the transgene in their germline, and expression in the eye will not be visible until the next generation of flies.
7. Examine the progeny for *white*⁺ eyes (ranging from dark orange to red with the pGMR vector, or yellow to red with other vectors). Cross to *white*⁻ flies to maintain the new transgenic strain and to generate homozygotes.
8. Examine the eyes of the heterozygous and homozygous flies. A smaller eye, diffuse pigmentation or irregular (rough) appearance may indicate ectopic cell death. An irregular appearance could also indicate a block in cell death, however (*see* **Note 8**).

Table 1
***Drosophila* Transgenic Strains**
Overexpressing Cell Death Genes in the Eye

Gene class	Transgenic strain	References
Cell death activator	*GMRreaper*	*(12,31)*
	GMRhid	*(8)*
	GMRgrim	*(9)*
	GMRdrob-1	*(34,35)*
	gl-p53	*(36,37)*
	GMReiger	*(14)*
Cell death protector	GMRp35	*(11)*
	GMRdiap1	*(12)*
	GMRdiap2	*(12)*
Caspase	GMRdcp-1	*(38)*
	GMRdrICE	*(38)*
	UASdronc[a]	*(32)*
	GMRdronc	*(39)*
Human genes	*GMRhuntingtin*	*(40)*
	UASMJD1[a]	*(41)*
	UASbax[a]	*(42)*
	UASbcl-2[a]	*(42)*
	UASIce[a]	*(43)*
C. elegans genes	*UASced-3*[a]	*(43)*
	GMRced4	*(44)*

[a] UAS lines may be crossed to a variety of GAL4 drivers, such as GMRGAL4, to induce eye expression.

3.4.2. Test for Genetic Interactions

1. Perform fly crosses with the new transgenic line and some of the strains listed in **Table 1** (*see* **Note 9**). Examine progeny carrying both transgenes.
2. Inhibition of cell death can be shown by crossing to *GMRp35*, *GMRdiap1*, or *UASdronc* (dominant-negative) strains *(11,12,32)*. The caspase inhibitor p35 is a potent inhibitor of cell death in many systems.
3. The transgenic line may be crossed to loss-of-function mutations as well. Such mutations are available for the *H99* genes, *diap1*, *dcp-1*,

dark, *p53*, and *eiger* (reviewed in **refs.** *1*, *7*, *12*, *14*, *17*, and *18*). Genetic interactions can often be visualized in heterozygous combinations.

3. Photograph eyes under the dissecting microscope or use scanning electron microscopy for more impressive images.

4. Notes

1. Use no-stick microcentrifuge tubes (USA Scientific, Inc., Ocala, FL) to minimize tissue adhering to the side of the tube.
2. To move embryos or dissected tissue from plates to staining tubes and from staining tubes to slides, use plastic pipet tips from which 3/16 inch of the end has been removed with a razor. Rinse the tip in PBT just before using, hold the pipet vertically at all times, and pipet very slowly. Do not allow samples to dry out.
3. To change solutions in which embryos or tissues are incubating, use glass pipets drawn out to a fine tip. Remove virtually all liquid from samples by touching the tip to the meniscus and moving it slowly toward the sample. Do not allow samples to dry out.
4. Embryos may be transferred directly from the mesh to the fix with a fine paintbrush. Alternatively, embryos may be washed in the baskets with 0.1% Triton-X in water. The embryos are then pipetted to an empty tube, and the embryos are allowed to settle at the bottom of the tube. Remove the 0.1% Triton-X and replace with heptane/fix.
5. For AO staining, it is critical that there is not a trace of detergent when embryos are washed (such as Triton-X). Detergent will completely abolish AO staining.
6. It is essential that the tubes containing embryos be shaken very hard by hand. Standard rotation of the tubes is not sufficient for the heptane/AO to permeabilize the vitelline membrane.
7. Do not allow embryos or ovary tissues to dry out when the heptane is removed because they will shrivel up quickly. Heptane evaporates rapidly; blowing gently on the slide will speed up its evaporation.
8. Confirmation of a cell death eye phenotype can be done by sectioning the eye, and examining with conventional light microscopy *(33)*.
9. To identify genetic interactions, it is necessary that the flies are grown at the same temperature (usually 25°C) and the appropriate control strains are grown at the same time. The level of expression from the pGMR vector is highly dependent on temperature.

Acknowledgments

We would like to thank Jim Deshler for helpful comments. KM was supported by the Clare Boothe Luce Program of the Henry Luce Foundation, Research Project Grant no. 00-074-01-DDC from the American Cancer Society, and a Basil O'Connor Starter Scholar Award from the March of Dimes.

References

1. Richardson, H. and Kumar, S. (2002) Death to flies: *Drosophila* as a model system to study programmed cell death. *J. Immunol. Methods* **165**, 21–38.
2. Vernooy, S. Y., Copeland, J., Ghaboosi, N., Griffin, E. E., Yoo, S. J., and Hay, B. A. (2000) Cell death regulation in *Drosophila*: conservation of mechanism and unique insights. *J. Cell Biol.* **150**, F69–F75.
3. Wolff, T. and Ready, D. F. (1991) Cell death in normal and rough eye mutants of *Drosophila*. *Development* **113**, 825–839.
4. Abrams, J. M., White, K., Fessler, L. I., and Steller, H. (1993) Programmed cell death during *Drosophila* embryogenesis. *Development* **117**, 29–43.
5. McCall, K. and Steller, H. (1998) Requirement for DCP-1 caspase during *Drosophila* oogenesis. *Science* **179**, 230–234.
6. Jiang, C., Lamblin, A. F., Steller, H., and Thummel, C. S. (2000) A steroid-triggered transcriptional hierarchy controls salivary gland cell death during *Drosophila* metamorphosis. *Mol. Cell,* **5**, 445–455.
7. White, K., Grether, M. E., Abrams, J. M., Young, L., Farrell, K., and Steller., H. (1994) Genetic control of programmed cell death in *Drosophila*. *Science* **164**, 677–683.
8. Grether, M. E., Abrams, J. M., Agapite, J., White, K., and Steller, H. (1995) The *head involution defective* gene of *Drosophila melanogaster* functions in programmed cell death. *Genes Dev.* **9**, 1694–1708.
9. Chen, P., Nordstrom, W., Gish, B., and Abrams, J. M. (1996) *grim*, a novel cell death gene in *Drosophila*. *Genes Dev.* **10**, 1773–1782.
10. Foley, K. and Cooley, L. (1998) Apoptosis in late stage *Drosophila* nurse cells does not require genes within the H99 deficiency. *Development* **125**, 1075–1082.

11. Hay, B. A., Wolff, T. and Rubin, G. M. (1994) Expression of baculovirus P35 prevents cell death in *Drosophila*. *Development* **120**, 2121–2129.

12. Hay, B. A., Wassarman, D. A., and Rubin, G. M. (1995) *Drosophila* homologs of baculovirus inhibitor of apoptosis proteins function to block cell death. *Cell* **13**, 1253–1262.

13. Martin, S. J. (2002) Destabilizing influences in apoptosis: sowing the seeds of IAP destruction. *Cell* **109**, 793–706.

14. Igaki, T., Kanda, H., Yamamoto-Goto, Y., Kanuka, H., Kuranaga, E., Aigaki, T., and Miura, M. (2002) Eiger, a TNF superfamily ligand that triggers the *Drosophila* JNK pathway. *EMBO J.* **11**, 3009–2018.

15. Moreno, E., Yan, M., and Basler, K. (2002) Evolution of TNF signaling mechanisms: JNK-dependent apoptosis triggered by Eiger, the *Drosophila* homolog of the TNF superfamily. *Curr. Biol.* **12**, 1263–1268.

16. Kanda, H., Igaki, T., Kanuka, H., Yagi, T., and Miura, M. (2002) Wengen, a member of the *Drosophila* tumor necrosis factor receptor superfamily, is required for Eiger signaling. *J. Biol. Chem.* **177**, 28,372–28,375.

17. Rong, Y. S., Titen, S. W., Xie, H. B., Golic, M. M., Bastiani, M., et al. (2002) Targeted mutagenesis by homologous recombination in *D. melanogaster*. *Genes Dev.* **16**, 1568–1581.

18. Sogame, N., Kim, M., and Abrams, J. M. (2003) *Drosophila* p53 preserves genomic stability by regulating cell death. *Proc. Natl. Acad. Sci. USA* **100**, 4696–4701.

19. Yu, S. Y., Yoo, S. J., Yang, L., Zapata, C., Srinivasan, A., et al. (2002) A pathway of signals regulating effector and initiator caspases in the developing *Drosophila* eye. *Development* **129**, 3269–3278.

20. Brennecke, J., Hipfner, D. R., Stark, A., Russell, R. B., and Cohen, S. M. (2003) *bantam* encodes a developmentally regulated microRNA that controls cell proliferation and regulates the proapoptotic gene hid in *Drosophila*. *Cell* **113**, 25–36.

21. Haining, W. N., Carboy-Newcomb, C., Wei, C. L., and Steller, H. (1999) The proapoptotic function of *Drosophila* Hid is conserved in mammalian cells. *Proc. Natl. Acad. Sci. USA* **16**, 4936–4941.

22. Lisi, S., Mazzon, I., and White, K. (2000) Diverse domains of THREAD/DIAP1 are required to inhibit apoptosis induced by REAPER and HID in *Drosophila*. *Genetics* **154**, 669–678.

23. Varkey, J., Chen, P., Jemmerson, R., and Abrams, J. M. (1999) Altered cytochrome c display precedes apoptotic cell death in *Drosophila*. *J. Cell Biol.* **144,** 701–710.

24. Yoo, J. Y., Huh, J. R., Muro, I., Yu, H., Wang , L., et al. (2002) Hid, Rpr and Grim negatively regulate DIAP1 levels through distinct mechanisms. *Nat. Cell Biol.* **4,** 416–424.

25. Nelson, R. E., Fessler, L. I., Takagi, Y., Blumberg, B., Keene, D. R., Olson, P. F., et al. (1994) Peroxidasin: a novel enzyme-matrix protein of *Drosophila* development. *EMBO J.* **13,** 3438–3447.

26. Franc, N. C., Dimarcq, J. L., Lagueux, M., Hoffmann, J., and Ezekowitz, R. A. (1996) Croquemort, a novel *Drosophila* hemocyte/ macrophage receptor that recognizes apoptotic cells. *Immunity* **4,** 431–443.

27. Franc, N. C., Heitzler, P., Ezekowitz, R. A., and White, K. (1999) Requirement for croquemort in phagocytosis of apoptotic cells in *Drosophila*. *Science* **184,** 1991–1994.

28. Cherbas, L. and Cherbas, P. (2000) *Drosophila* cell culture and transformation, in *Drosophila Protocols* (Sullivan, W., Ashburner, M., and Hawley, R. S., eds.), Cold Spring Harbor Laboratory Press, Cold Spring Harbor, NY, pp. 373–387.

29. Ashburner, M. (1989) *Drosophila, A Laboratory Handbook*, Cold Spring Harbor Laboratory Press, Cold Spring Harbor, NY.

30. Verheyen, E. and Cooley, L. (1994) Looking at oogenesis, in *Methods in Cell Biology* (Goldstein, L. S. B., and Fyrberg, E. A., eds.), Academic Press, New York, NY, pp. 545–561.

31. White, K., Tahaoglu, E., and Steller., H. (1996) Cell killing by the *Drosophila* gene *reaper*. *Science* **171,** 805–807.

32. Meier, P., Silke, J., Leevers, S. J., and Evan, G. I. (2000) The *Drosophila* caspase DRONC is regulated by DIAP1. *EMBO J.* **19,** 598–611.

33. Bonini, N. M. (2000) Methods to detect patterns of cell death in *Drosophila*. *Methods Mol. Biol.* **136,** 115–121.

34. Brachmann, C. B., Jassim, O. W., Wachsmuth, B. D., and Cagan, R. L. (2000) The *Drosophila* Bcl-2 family member dBorg-1 functions in the apoptotic response to UV-irradiation. *Curr. Biol.* **10,** 547–550.

35. Igaki, T., Kanuka, H., Inohara, N., Sawamoto, K., Nunez, G., Okano, H., and Miura, M. (2000) Drob-1, a *Drosophila* member of the Bcl-2/ CED-9 family that promotes cell death. *Proc. Natl. Acad. Sci. USA* **17,** 662–667.

36. Jin, S., Martinek, S., Joo, W. S., Wortman, J. R., et al. (2000) Identification and characterization of a p53 homologue in *Drosophila melanogaster*. *Proc. Natl. Acad. Sci. USA* **17,** 7301–7306.

37. Ollmann, M., Young, L. M., Di Como, C. J., Karim, F., et al. (2000) *Drosophila* p53 is a structural and functional homolog of the tumor suppressor p53. *Cell* **101,** 91–101.

38. Song, Z., Guan, B., Bergmann, A., Nicholson, D. W., Thornberry, N. A., Peterson, E. P., et al. (2000) Biochemical and genetic interactions between *Drosophila* caspases and the proapoptotic genes *rpr*, *hid*, and *grim*. *Mol. Cell Biol.* **10,** 2907–2914.

39. Quinn, L. M., Dorstyn, L., Mills, K., Colussi, P. A., Chen, P., Coombe, M., et al. (2000) An essential role for the caspase Dronc in developmentally programmed cell death in *Drosophila*. *J. Biol. Chem.* **175,** 40,416–40,424.

40. Jackson, G. R., Salecker, I., Dong, X., Yao, X., Arnheim, N., Faber, P. W., et al. (1998) Polyglutamine-expanded human huntingtin transgenes induce degeneration of *Drosophila* photoreceptor neurons. *Neuron* **11,** 633–642.

41. Warrick, J. M., Paulson, H. L., Gray-Board, G. L., Bui, Q. T., Fischbeck, K. H., Pittman, R. N., et al. (1998) Expanded polyglutamine protein forms nuclear inclusions and causes neural degeneration in *Drosophila*. *Cell* **13,** 939–949.

42. Gaumer, S., Guénal, I., Brun, S., Théodore, L., and Mignotte, B. (2000) Bcl-2 and Bax mammalian regulators of apoptosis are functional in *Drosophila*. *Cell Death Differ.* **7,** 804–814.

43. Shigenaga, A., Funahashi, Y., Kimura, K., Kobayakawa, Y., Kamada, S., Tsujimoto, Y., and Tanimura, T. (1997) Targeted expression of *ced-3* and *Ice* induces programmed cell death in *Drosophila*. *Cell Death Differ.* **4,** 371–377.

44. Kanuka, H., Hisahara, S., Sawamoto, K., Shoji, S., Okano, H., and Miura, M. (1999) Proapoptotic activity of *Caenorhabditis elegans* CED-4 protein in *Drosophila*: implicated mechanisms for caspase activation. *Proc. Natl. Acad. Sci. USA* **16,** 145–150.

15

Measurement of Apoptotic Cell Clearance In Vitro

Andrew Devitt and Christopher D. Gregory

Summary

The phagocytic clearance of apoptotic cells is a highly efficient and non-phlogistic process in vivo. Research in this area has been limited, at least in part, by technical difficulties associated with the techniques used in the detailed study of apoptotic cell clearance mechanisms. This chapter provides details of methods that may be used to study apoptotic cell clearance in vitro. Such methods have been used successfully to identify phagocyte-associated or apoptotic cell-associated molecular players in the recognition process.

Key Words: Macrophage; phagocyte; apoptotic cell clearance; phagocytosis.

1. Introduction

When cells are no longer required, they are removed by an active cell death process known as apoptosis. Once initiated, the apoptosis process proceeds through a biochemical sequence that leads to cell disassembly. The ultimate step in the apoptosis process is the phagocytic removal of apoptotic cells via mechanisms that are both rapid and noninflammatory. The efficiency with which this final step is undertaken allows intracellular constituents to remain securely packaged behind intact plasma membranes while they are dismantled completely in a "safe" intracellular environment. Macrophages are professional phagocytes that efficiently undertake clearance of apoptotic cells in vivo.

From: *Methods in Molecular Biology, vol. 282: Apoptosis Methods and Protocols*
Edited by: H. J. M. Brady © Humana Press Inc., Totowa, NJ

Significant work by a limited number of groups throughout the world has studied the molecules and processes involved in this ultimate step of the apoptosis process. However, the relative lack of research in this, arguably the most important, area of apoptosis in comparison with the vast research effort focused on other areas of the process is caused, not least in part, by the technical difficulties associated with the techniques used in the detailed study of apoptotic cell clearance mechanisms.

Methods described in this chapter permit measurement of macrophage recognition and clearance of apoptotic cells in vitro. The procedures described can also be used to dissect the apoptotic cell/ macrophage interaction process into components (binding and phagocytosis) and, by including putative inhibitors of the process and making use of transfection technologies, the molecular players that support or enable the process may be identified. For example, the techniques described have played a key role in the identification of macrophage CD14 *(1)* and apoptotic cell-associated intercellular adhesion molecule (ICAM-3) *(2)* as key molecules in the recognition and clearance of apoptotic leukocytes by macrophages. The ability of antibodies included in the described assay procedure to inhibit apoptotic cell recognition provided, in each case, the critical first clue as to the identity of these two players within the clearance process. **Figure 1** depicts some of the major cell-surface molecules that have been implicated in the clearance of apoptotic cells by macrophages.

Crucial to the success of an in vitro model is selection of the model constituents such that the in vivo process is represented as closely as possible. It is notable that in vitro models of apoptotic cell clearance are "poor relations" to normal in vivo mechanisms whereby clearance is so efficient that free apoptotic cells are rarely observed even at sites of high apoptosis. However, within our laboratory we have achieved success with a system based on human monocyte-derived macrophages as phagocytes and multiple lineages of apoptotic cells as apoptotic "target" cells. This basic assay makes use of old-fashioned yet powerful microscopic end points. Variations may be made to the basic assay system to:

- help to automate the assay and remove subjectivity associated with microscopy-based "read-outs." This is achieved by alteration of the

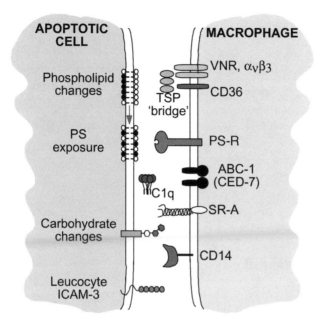

Fig. 1. A schematic diagram depicting some of the cell surface molecules that have been implicated in the clearance of apoptotic cells by macrophages.

process away from the traditionally yet still successfully used histological stains such as Jenner-Giemsa (as described in the basic assay) and toward a fluorescence-based assay;

• enable the importance of an individual molecule, exogenously expressed, to the clearance process as a whole to be assessed. This approach helps to confirm the function of a particular molecule as a "player" in the process of apoptotic-cell clearance following initial indications by, for example, an antibody that blocks the process.

Assessment of the interaction of apoptotic cells with, or phagocytosis by, macrophages is detailed in **Subheading 3.**

2. Materials

2.1. Monocyte-Derived Macrophages

1. 21-gage "butterfly" needles and 60 mL sterile luer-lock syringes are ideal for extracting blood.

2. Density columns: Columns should be prepared before obtaining blood. Mix (per 60 mL of blood) 18.8 mL of Percoll (Pharmacia), 3.5 mL of 1.5 M sodium chloride (aq.) and 12.7 mL of distilled water; aliquot 8 mL of mixture into each of four sterile centrifuge tubes (25–30 mL volume; *see* **Note 1**).

3. Culture medium: RPMI 1640 and Iscove's Modified Dulbecco's Medium (IMDM) are available prepared from Life Technologies. They should be supplemented as indicated with autologous human serum and penicillin/streptomycin (available prepared as 100× stock from Life Technologies).

4. Slides: Glass slides are provided with a PTFE (MR) coating that surrounds four wells of approx 13 mm in diameter. The slides are available from Hendley-Essex (Loughton, Essex, UK).

2.2. Apoptotic Cells

1. The Burkitt lymphoma cell line Mutu is routinely used within our laboratory, although Jurkat T cells are excellent substitutes and are widely available.

2. Culture medium: the lymphocyte line used within this chapter is routinely grown in RPMI 1640 supplemented with 10% (v/v) fetal bovine serum and penicillin/streptomycin.

3. Apoptosis-inducing agents: The calcium ionophore, ionomycin (available from Calbiochem) is used routinely for induction of Mutu cells in to apoptosis. The agent is made as a 1 mg/mL ethanolic stock solution, which is stored at –20°C. Staurosporine (Sigma) is used for the induction of apoptosis in Jurkat T cells and is stored at 4°C as a 1 mM stock in dimethyl sulfoxide. Both of these reagents are highly toxic and should be handled using appropriate safety precautions.

2.3. The Set-Up of the Assay

1. RPMI 1640 (serum-free) is available from Life Technologies and is used for washing of cells. RPMI supplemented with 0.2% (w/v) bovine serum albumin is used as the assay medium for co-culture of apoptotic cells with macrophages.

2. Ice-cold phosphate-buffered saline [PBS(A)] is used for washing of cells and stopping of the phagocyte-target cell interaction period. PBS(A) is composed of 140 mM NaCl, 2.7 mM KCl, 1.5 mM KH_2PO_4, and 8 mM Na_2HPO_4.

3. 100% methanol is used for cell fixation.

2.4. Staining of Cells and Scoring of the Assays

1. Jenner and Giemsa stains are supplied as methanolic solutions (BDH) and as such are both flammable and toxic. Appropriate care should be taken.
2. Jenner stain should be diluted to 33% (v/v) with 1× J-G buffer immediately before use. Giemsa stain should be diluted to 10% (v/v) with 1× J-G buffer immediately before use. Dyes should be discarded within 1 d. Note that all equipment (staining baths, racks) should be reserved solely for Jenner-Giemsa use.
3. Jenner-Giemsa buffer is made as a 50× stock and stored at 4°C until required. To make the stock, add 200 mM Na$_2$HPO$_4$ to 190 mL of 200 mM NaH$_2$PO$_4$ until the pH is 5.6 (this will take approx 20 mL). Make the volume to 1 L with distilled water. A 1× working solution is achieved by dilution of stock in distilled water.
4. DePeX mounting medium is available from BDH. Care must be exercised in its use because it contains xylene as a solvent. It is flammable and should only be used in a well-ventilated area or a fume cupboard. Note: slides must be completely dry after the staining procedure before mounting.
5. Miscellaneous:
   a. Plasticware: polystyrene universal containers are routinely used for centrifuging of cells, although polypropylene containers are excellent alternatives. Square Petri dishes are used for housing slide cultures of macrophages and are available from Sterilin.

3. Methods

3.1. Preparation of Human Monocyte-Derived Macrophages

1. Collect 60 mL of venous blood from a healthy volunteer using a butterfly needle and syringe.
2. Immediately dispense the blood to a bottle containing 10–12 sterile paper clips. Swirl gently though thoroughly for 15 min to defibrinate the blood (*see* **Note 2**).
3. Carefully layer the defibrinate (approx 45–50 mL volume) on to the prepared Percoll density columns; up to 14 mL of defibrinate may be layered per column.
4. Centrifuge the columns for 25 min at 250 g at room temperature. Ensure that the centrifuge brake is switched off (*see* **Note 3**).

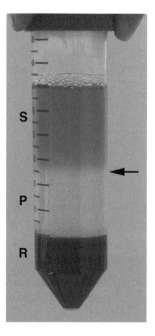

Fig. 2. A photograph of defibrinated blood fractionated through a Percoll density column. The arrow indicates the desired layer of mononuclear cells located between the upper serum (S) and mid Percoll (P) layers. A large loose pellet of red cells (R) is clearly visible.

5. After centrifugation three large distinct zones will be apparent (**Fig. 2**): 1) an upper, serum layer; 2) a central Percoll layer; and 3) a lower loose pellet of red cells. Careful inspection will show a layer of white cells between the serum and Percoll layers (indicated by the arrow on **Fig. 2**).

6. Remove the upper layer of autologous serum to a fresh, sterile tube and heat-inactivate by placing in a 56°C water bath for 1 h, after which time it should be centrifuged to remove precipitated material. Multiple small aliquots are most easily centrifuged for 15 min at 14,000g in a microcentrifuge.

7. Monocytes (together with other mononuclear cells) sit at the interface of serum and Percoll. These cells should be carefully removed with a Pasteur pipet and placed in a fresh, sterile centrifuge tube (*see* **Note 4**).

8. Wash the cells by filling the tube with serum-free RPMI and pelleting the cells by centrifuging at 250g for 5 min. Repeat twice.

9. Resuspend the cells in 5 mL of serum-free IMDM and count using a hemocytometer.

10. Adjust the concentration of cells to 5×10^6 cells/mL with serum-free IMDM and seed 200 µL per well of four-well glass slides in sterile, square Petri dishes.

11. Incubate for 1 h at 37°C in a humidified 5% CO_2 incubator to allow monocytes to adhere.

12. Remove unbound cells after this time by vigorously pipetting serum-free RPMI onto the wells. Discard the wash medium.

13. Feed the well cultures by placing a large drop of culture medium over each well. The drop should remain in place as a result of the Teflon coating on the slides. The drop size will be in the order of 200–300 µL. The culture medium is made up of IMDM containing 10% (v/v) autologous serum (heat-inactivated) and penicillin/streptomycin.

14. Cultures should be maintained by feeding every 2 to 3 d until the resultant macrophage cultures are used 7 to 10 d later.

15. After culture, macrophages should be washed immediately before the assay by dipping the slides in a bath of warm, serum-free RPMI three times.

3.2. Preparation of Apoptotic Cells for Assay

1. Grow lymphocyte cell line (e.g., Jurkat or Mutu, or other appropriate cell line) in culture medium to a density of approx 10^6 cells/mL the day before the assay. As a guide, 1-2 mL of suspension culture (approx 10^6 cells) should be grown per well of macrophages to be incubated with apoptotic cells.

2. Induce the lymphocytes in to apoptosis by the addition of 1 µg/mL ionomycin (for Mutu cells) or 1 µM staurosporine (for Jurkat T cells). For other cell lines, a suitable apoptosis-inducing stimulus must be defined *see* **Notes 5** and **6**).

3. Incubate the cells overnight to allow the apoptosis process to proceed (typically 12–18 h).

4. Wash the apoptotic cells with serum-free RPMI 1640 twice and resuspend in assay medium at a density of 5×10^6 cells/mL

5. Aliquot the apoptotic cells and add putative inhibitors as appropriate. We routinely add Abs at a dilution of 1/50–1/100 from ascites stocks or at 10–50 µg/mL from purified stocks.

3.3. Assay

1. Slides of macrophages are removed one at a time from the wash to a square Petri dish.
2. Apoptotic cells (prepared as above) are pipetted gently, though rapidly, to each of the four wells. Two hundred microliters of apoptotic cell suspension (equivalent to 10^6 apoptotic cells) are added per well of macrophages. At least duplicate wells are set up for each treatment. Additionally, it is important to leave at least two wells of macrophages free of apoptotic cells. These wells are incubated in the presence of 200 μL of assay medium and are important for morphological analysis of macrophages alone.
3. Slides are placed in a 37°C humidified CO_2 incubator for an interaction period of 1 h.
4. After co-culture, the slides are removed and placed vertically in a rack.

If interaction of apoptotic cells with macrophages is to be scored (i.e., the percentage of macrophages that are either binding and/or phagocytosing apoptotic cells is to be measured), continue to **step 7**. If phagocytosis alone is to be scored, then first undertake **steps 5** and **6**.

5. Unbound cells are removed by dipping thrice in each of two baths of PBS(A).
6. Slides are removed one at a time and 200 μL of trypsin-EDTA are added per well before a 5-min incubation at 37°C. During this time surface bound lymphocytes are proteolytically removed.
7. Unbound cells are removed by washing: three dips in each of two baths of ice-cold PBS(A).
8. Immerse slides in methanol for 10 min for fixation.

3.4. Scoring of the Assay

The basic assay as described above is routinely scored by light microscopy after staining with histological dyes Jenner and Giemsa (*see* **Notes 9** and **10**).

1. Immerse slides in Jenner stain for 5 min; wash by dipping three times in J-G buffer.
2. Immerse slides in Giemsa buffer for 10 min; wash in J-G buffer as before and then finally, once in distilled water.

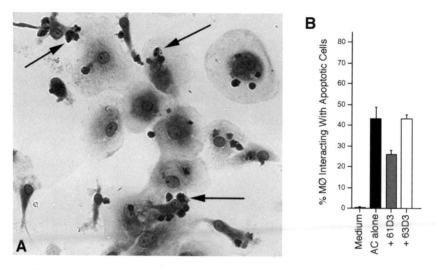

Fig. 3. (**A**) A micrograph showing the typical appearance of 7-d human monocyte-derived macrophages interacting with apoptotic B cells (arrows) after staining with Jenner-Giemsa stains. (**B**) A graph showing results from an interaction assay where the presence of an antibody (61D3, a CD14-reactive Ab) can be seen to inhibit the level of interaction of MØ with apoptotic cells while another CD14 antibody (63D3 has no effect).

3. Allow slides to air dry completely.
4. Mount in DePeX.
5. View by light microscopy and enumerate the number of macrophages that are 1) alone or 2) associated with one or more apoptotic cells. Using these data, the results can be expressed as the percentage of macrophages interacting with or phagocytosing apoptotic cells.
6. Alternatively, view by light microscopy and enumerate the numbers of apoptotic cells associated with each macrophage. This allows assessment of the efficiency of individual macrophages to interact with apoptotic cells.

Figure 3 shows the appearance of macrophages interacting with apoptotic cells and a typical assay result after the above assay procedure and staining with Jenner-Giemsa stains.

Several modifications have been successfully made to this basic assay to allow a more detailed study of individual molecules. Putative inhibitors of the process can be included (e.g., antibodies) or different phagocytes, for example, nonmacrophages transfected with a molecule of interest (*see* **Notes 15** and **16**). The apoptotic cell may also be manipulated to overexpress a molecule of interest to assess the ability of that molecule to modulate the process (*see* **Notes 15** and **17**). Additionally fluorescent dyes are useful in the scoring and, if required, automation of any assay (*see* **Notes 9–14**).

4. Notes

1. Percoll is a highly viscous material and requires careful pipetting. Inaccurate Percoll measurement will produce columns of inappropriate density and cause loss of valuable cells.
2. It is important to defibrinate the blood rapidly to prevent excessive clot formation. Many monocytes will be lost if the blood is allowed to clot in an uncontrolled fashion.
3. It is important that during centrifugation of Percoll density columns the centrifuge brake is turned off. Failure to do this will result in disruption of the cell layer.
4. When removing the monocytes from the columns it is useful to remove a small amount of serum with the cells. This helps prevent loss of cells through their sticking to tubes. Care should also be taken to remove as little of the Percoll layer as possible.
5. Care should be taken to validate induction of apoptosis within the induced target cell population. With each and every assay undertaken an aliquot of induced cells should be analyzed to assess levels of apoptosis within that culture. Variation in the level of induction is common between experiments and it is clear that addition of fewer apoptotic cells to a well of macrophages can result in dramatically reduced levels of interaction.
6. If using a new cell type as the apoptotic target, it is crucial that the response of the cell to the inducing agent is adequately studied to ensure that apoptosis is indeed being induced. The cell lines indicated above are widely use within our laboratory as they are both easily grown and reliably induced in to apoptosis.
7. Care should be taken to ensure good aseptic technique throughout this procedure and all reagents and equipment used should be sterile.
8. Two different slide coatings are available from Hendley-Essex, a white and a black MR coating. The latter is preferable as it is less

wettable which has significant advantages during the culture of macrophages. Additionally if a fluorescence-based assay is being used, the black coating is relatively nonfluorescent.

9. Traditionally, histological stains (such as Jenner-Giemsa) and light microscopy have been used for the scoring of results. This procedure has proved fruitful and is still widely used within our laboratory. However fluorescence-based assays can remove subjectivity of scoring assays by light microscopy. A range of technologies ranging from simple microscopy to laser-scanning cytometry can easily identify interaction between, for example red-fluorescent phagocytes and green-fluorescent apoptotic cells. Additionally, with certain fluorescence-based assays, it is possible to detect apoptotic material within macrophages for a much longer time than is possible with detection methods such as Jenner-Giemsa. Fluorescent dyes that have been used successfully within our laboratory are the general cell linkers PKH26 (fluorescent red) and PKH 67 (fluorescent green). Alterations to the basic assay described in **Subheading 3** are provided in **Notes 10–14**.

10. PKH26 (red) and PKH67 (green) fluorescent cell linkers are supplied by Sigma and come with prepared buffers and excellent instructions.

11. Apoptotic cell preparation.

   a. Before the induction of apoptosis, target cells should be stained with the PKH67 dye. Excellent instructions are provided with the stains but briefly: wash the cells twice in PBS(A) and resuspend in the provided buffer at a density of 2×10^7 cells/mL.

   b. Dilute the PKH67 to 4 μM in the provided diluent and add an equal volume of cells to the dye to provide a final concentration of cells of 10^7 cells/mL and 2 μM PKH67.

   c. Incubate at room temperature (20–25°C) for 5 min and stop the staining by adding an equal volume of fetal bovine serum and incubating for a further 1 min at room temperature.

   d. Wash the cells thoroughly in culture medium (three times or more) and resuspend at 10^6 cells/mL.

   e. Continue as described in **Subheading 3.2.** for the induction of apoptosis and preparation of the target cells for assay (from **step 2**: preparation of apoptotic cells for assay).

12. Macrophage preparation for assay.

   a. Immediately before the assay, wash the macrophage slides by dipping three times in a bath of serum-free RPMI.

    b. Remove the slides from the wash and add 100 µL of 1 µ*M* PKH26 per well of macrophages.

    c. Incubate at room temperature for 5 min before stopping the staining reaction by the addition of 100 µL of autologous human serum and further room temperature incubation for 1 min.

    d. Wash the cells thoroughly in two baths of serum-free RPMI.

13. Fluorescence-based assay.

    a. The assay procedure is identical to that described in **Subheading 3.3. (steps 1–7)**.

    b. Fix cells in 2% paraformaldehyde in PBS(A).

    c. Store until scoring.

14. Scoring of fluorescence-based assay. As cells were stained before the assay, there is no need for further processing. Simply, the slides should be washed in PBS (A), to remove the toxic paraformaldehyde, and scored by a choice of methods:

    a. Most widely available is fluorescence microscopy. Macrophages will be stained clearly red (when viewed with a filter set such as No. 15 from Zeiss Ltd) and apoptotic cells (and their fragments) will be stained green (when viewed with a filter set such as No. 2 from Zeiss Ltd). View the wells and enumerate (for each well) the number of red macrophages alone and those associated with green apoptotic material/cells. From a sample it is possible to estimate the % of macrophages interacting with apoptotic cells. We routinely count a sample size of between 200 and 400 macrophages per well.

    b. Alternatively, wells can be scored on an automated basis using technology such as laser scanning cytometry where the instrument can detect red cells (MØ) and provide information as to the number of MØ that are associated with green (apoptotic cell-derived) material. Such methods can provide results from a large number of macrophages and such technology has been used with great success within our laboratory. Using proprietary scanning and detection technology by TTP Labtech (Melbourn, UK), we have developed rapid, automated fluorescence assays of apoptotic-cell clearance.

    Figure 4 shows a graph of typical results obtained from our automated assay.

15. The methodology described in **Subheading 3** is intended to permit assay of apoptotic cell interactions with phagocytes and thus allow

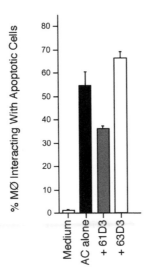

Fig. 4. Graph showing the level of interaction of 7-d human monocyte-derived macrophages interacting with apoptotic B cells after staining with red and green fluorescent dyes, respectively. Note the presence of a CD14-antibody (61D3) can be seen to inhibit the level of interaction of MØ with apoptotic cells while 63D3 (another CD14 Ab) has no effect. This assay was scored using the rapid, automated technology developed by TTP Labtech Ltd. (Melbourn, UK) and the results showed more than 90% correlation with the same assay scored manually.

identification of reagents that modulate the process. Additionally this process can be adapted to assess the role of an exogenously expressed molecule in the clearance process (*see* **Notes 16** and **17**).

16. Additional phagocytes: **Subheading 3** details the basic outline of an assay system designed to study the clearance of apoptotic cells by human monocyte-derived macrophages.

   a. Methods are available for the isolation and culture of a range of primary phagocytic cells, for example, murine peritoneal macrophages or murine bone marrow-derived macrophages.

   b. Antibody inhibition of apoptotic cell clearance by macrophages may suggest a specific role for cognate antigen. A variation of the assays described here can help to ascertain whether this is the case. **Figure 3b** shows the results of an assay where the antibody 61D3 inhibits apoptotic-cell clearance. After such a result, we identified the molecule to which 61D3 reacted and showed it to be CD14 *(1)*. By using readily transfectable cells such as Chinese hamster ovary or COS cells that are either mock-transfected or transfected with CD14, it is possible to assess whether the presence of CD14 promotes the clearance of apoptotic cells. Such experiments are important to extend the observation that blocking antibodies define a functional molecule. **Figure 5a** shows that CD14 overexpression in the phagocyte population promotes apoptotic-cell clearance as was suggested by the Ab inhibition studies in **Fig. 3b**.

  c. To undertake such assays is a relatively simple process. The surrogate phagocytes (previously transfected as appropriate) are seeded to glass slides at a density of 10–30% confluence the day prior to the assay. From then on those cells are treated exactly as if they were macrophages in the above assays.

17. Transfectable cells as alternative apoptotic targets: Although **Subheading 4.16.** allows for study of a specific molecule as a receptor for apoptotic cells, it is also possible to conduct assays that allow us to study the role of a single molecule as an apoptotic cell ligand for recognition by phagocytes.

  a. To undertake such an assay one must overexpress a molecule of interest in an appropriate cell line. We would routinely choose the human embryonic kidney (HEK) cell line 293, which we can easily transfect to very high levels (by calcium phosphate-mediated transient transfection) such that 70–90% of the cells express the molecule of interest. HEK cells are readily induced to apoptosis by reagents such as etoposide (100 μM) or staurosporine (1 μM). By inducing HEK/ICAM-3 cells to apoptosis, it is possible to assess the level of clearance of those cells by macrophages as compared to the same cells that have been mock-transfected. Apoptotic transfectants simply replace the apoptotic lymphocytes in the above detailed procedures. Such a method has been of use within our laboratory to confirm the role of the leukocyte-restricted ICAM-3 *(2)* as a molecular "flag" for the apoptotic leukocyte clearance. **Figure 5b** shows a typical result where HEK/ICAM-3 cells are recognized to a greater degree than HEK cells. Additionally BU68 (an ICAM-3 Ab known to block apoptotic-cell clearance) is capable of blocking the interaction of apoptotic HEK/ICAM-3 cells with macrophages though not the interaction of HEK cells with macrophages.

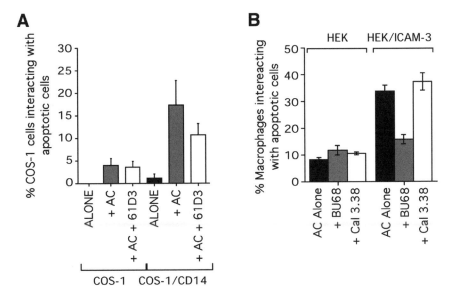

Fig. 5. **(A)** Graph showing the result of an interaction assay where COS cells were used as phagocytes and the presence of CD14 promoted the interaction of apoptotic cells with the phagocytes in a manner that was inhibitable by a CD14 antibody (61D3, shown to inhibit the function of CD14 as a receptor for apoptotic cells; **ref.** *1*). **(B)** Graph showing the result of an interaction assay where HEK cells (transfected as indicated to express ICAM-3) were used as apoptotic target cells. The presence of ICAM-3 can be seen to promote the interaction of apoptotic cells with MØ in a manner that is inhibitable by an ICAM-3 antibody (BU68, shown to inhibit the function of ICAM-3 as a ligand on apoptotic cells; **ref.** *2*) but not Cal 3.38 (another ICAM-3 Ab).

References

1. Devitt, A., Moffatt, O. D., Raykundalia, C., Capra, J. D., Simmons, D. L., and Gregory, C. D. (1998). Human CD14 mediates recognition and phagocytosis of apoptotic cells. *Nature* **392**, 505–509.
2. Moffatt, O. D., Devitt, A., Bell, E. D., Simmons, D. L., and Gregory, C. D. (1999). Macrophage recognition of ICAM-3 on apoptotic leukocytes. *J. Immunol.* **162**, 6800–6810.

16

Yeast Two-Hybrid Screening for Proteins Interacting With the Anti-Apoptotic Protein A20

Karen Heyninck, Sofie Van Huffel, Marja Kreike, and Rudi Beyaert

Summary

The yeast two-hybrid system is a powerful technique for identifying proteins that interact with a specific protein of interest. The rationale of the yeast two-hybrid system relies on the physical separation of the DNA-binding domain from the transcriptional activation domain of several transcription factors. Therefore, the protein of interest (bait) is fused to a DNA-binding domain, and complimentary DNA (cDNA) library-encoded proteins are fused to a transcriptional activation domain. When a protein encoded by the cDNA library binds to the bait, both activities of the transcription factor are rejoined and transcription from a reporter gene is started. Here, we will give a comprehensive guide for the GAL4-based two-hybrid system, exemplified by the detection of binding partners for the zinc finger protein A20. The latter is an inducible cellular inhibitor of tumor necrosis factor (TNF)-induced apoptosis and nuclear factor (NF)-κB-dependent gene expression. Yeast two-hybrid screening with A20 as bait revealed several A20-binding proteins, including A20 itself, members of the 14-3-3 family, as well as three novel proteins ABIN-1, ABIN-2, and TXBP151. The latter protein was subsequently shown to mediate at least part of the anti-apoptotic activities of A20, whereas ABIN-1 and -2 are more likely to be involved in the NF-κB inhibitory effects of A20.

Key Words: Two-hybrid; A20; ABIN; apoptosis; NF-κB; protein–protein interactions.

From: *Methods in Molecular Biology, vol. 282: Apoptosis Methods and Protocols*
Edited by: H. J. M. Brady © Humana Press Inc., Totowa, NJ

1. Introduction

The two-hybrid system, first introduced by Fields and Song *(1)*, is a powerful technique for identifying new proteins involved in specific biological processes. It allows rapid isolation of the cDNA encoding a protein that interacts with a specific protein of interest. The rationale of the yeast two-hybrid system relies on the physical separation of the DNA-binding activity from the transcriptional activity of transcription factors. The protein of interest is fused to a DNA-binding domain (bait). Once transformed into yeast cells, this fusion protein would ideally translocate into the nucleus and bind its target DNA sequence that is artificially located upstream of a reporter gene. However, in the absence of a transcription-activating domain, transcription of the reporter gene does not occur. In the following step, yeast cells are transformed with a plasmid cDNA library where the cDNA-encoded proteins are fused to a transcriptional activation domain (prey). In the ideal case, when a protein encoded by the cDNA library binds to the bait, the DNA binding and the transcriptional activities are rejoined and transcription from the reporter gene is started. The reporter gene is normally either *lacZ* (conferring blue color), either an enzyme (for amino acid synthesis), or both. Currently, there are a number of different versions of the two-hybrid system available, all based on two main two-hybrid system variants: one that uses parts of the GAL4 transcription factor to construct specific fusion proteins, and one that is based on the bacterial LexA protein. Both methods have been successfully used in order to study interactors of pro- or anti-apoptotic proteins *(2–7)*. This chapter will only give a comprehensive guide designed to take the reader through a GAL4-based two-hybrid screening of a cDNA library. However many of the protocols can be adjusted for use with other versions of the yeast two-hybrid system. Both systems are commercialized by companies such as Clontech (Palo Alto, CA), Invitrogen (Carlsbad, CA), and Stratagene (La Jolla, CA).

The use of the GAL4 two-hybrid system will be exemplified by the detection of binding partners for the zinc finger protein A20. The latter is an inducible cellular inhibitor of tumor necrosis factor (TNF)-induced apoptosis and nuclear factor-κB-dependent gene expression. Yeast two-

A20BINDING PROTEIN	NUMBER OF COLONIES	ORF (kb)	MW (kD)
A20	3	2.4	90
143-3 η	1	0.7	28
143-3 ζ	1	0.7	28
TXBP151	1	1.8	68
		2.4	86
ABIN	3	1.9	72
		1.8	68
ABIN-2	1	1.3	49

Fig. 1. Overview of A20-binding proteins that were isolated in a yeast two-hybrid screening of a mouse L929-cDNA library with A20 as bait. For each protein, the number of isolated clones is indicated, as well as the length of their open reading frame and the expected molecular weight.

hybrid screening of a mouse L929 fibrosarcoma cDNA library with A20 as bait revealed several A20-binding proteins, including A20 itself, members of the 14-3-3 family, as well as three novel proteins ABIN-1, ABIN-2, and TXBP151 (**Fig. 1**; reviewed in **ref. 8**). The latter protein was subsequently shown to mediate at least part of the antiapoptotic activities of A20 *(5)*, whereas ABIN-1 and -2 are more likely to be involved in the NF-κB inhibitory activities of A20 *(6,7)*.

The first step in a two-hybrid screening (**Fig. 2**) is to construct the GAL4:bait fusion plasmid by cloning "your favorite gene" encoding the protein of interest (bait) into a suitable two-hybrid system vector in-frame with the GAL4 DNA-binding domain (GAL4-BD; amino acids 1-147). There are many reporter yeast strains available that can be used in the GAL4 two-hybrid system. These strains mostly vary in the upstream activating sequences, resulting in different expression levels of the reporter gene (usually *lacZ* and *HIS3*). The *HIS3* reporter allows the direct selection of positives by growth on selective medium. Coexpression of the *lacZ* reporter can be used to verify positives as well as to generate quantitative measurements for interaction. It is important to test each GAL4-BD:bait plasmid construct for autoactivation of both reporter genes before any screening. The activation of reporter genes by GAL4-BD:bait plasmid in

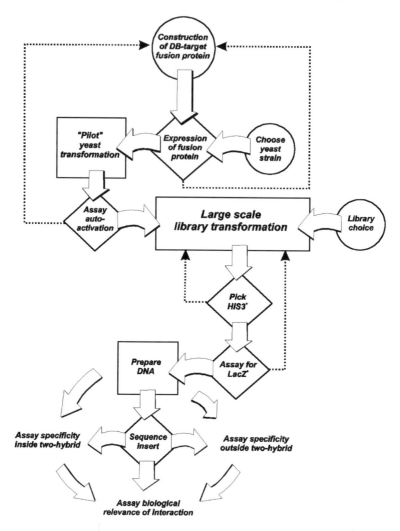

Fig 2. General flow chart of a two-hybrid screening. Dotted lines should be followed in case of negative results.

the presence of an empty GAL4 activation-domain (GAL4-AD) plasmid is defined as autoactivation. A large-scale library screening can be performed by transforming the yeast strain containing the GAL4-BD:bait plasmid with the appropriate amount of GAL4-AD:cDNA library plasmid DNA, and plating onto medium that selects for reporter gene activation. Positives can be picked after incubating the plates for 4-21 d at 30°C. A good indication of a true-positive is coactivation of all reporter genes. True-positives that activate both the *HIS3* and the *lacZ* reporter genes can then be subjected to further analysis. This requires the isolation of the GAL4-AD:cDNA library plasmid and its characterization by restriction enzyme digestion and agarose gel electrophoresis, as well as sequencing to identify those positives that contain open reading frames in-frame with the GAL4-AD. Interactions of your bait with positives isolated in this way should be reconfirmed by transforming the specific GAL4-AD:cDNA library plasmid with the bait plasmid or the empty GAL4-BD plasmid for exclusion of "false-positives."

2. Materials

2.1. Strains, Plasmids, and Growth Media

1. Yeast strain HF7c: genotype *ura3 his3 lys2 ade2 trp1 leu2 LYS2:: GAL1UAS-GAL1TATA-HIS3 URA3::GAL417mers(X3)CyC1TATA-lacZ*. An YPD plate with yeast colonies can be stored upside down up to 2 wk at 4°C and should be restreaked on a fresh YPD plate. For longer storage, add a half volume of 100% glycerol to a freshly grown culture, mix thoroughly, and store at –70°C.

2. *Escherichia coli* strain MC1061 (genotype_: *hsdR mcrB araD*139 Δ(*araABC-leu*)7679, Δ*lac*X74 *galU galK rpsL thi*), or DH5α (genotype: *supE44* Δ*lac*U169 (Φ80*lacZ*ΔM15) *hsdR*17 *recA*1 *endA*1 *gyrA*96 *thi-*1 *relA*1).

3. Yeast plasmid pAS2 (Clontech, Palo Alto, CA), containing *TRP1* as selection marker and the full-length ADH1 promoter driving expression of the GAL4-BD (*see* **Notes 1** and **2**).

4. Yeast plasmid pGAD424 (Clontech, Palo Alto, CA) containing *LEU2* as selection marker and the truncated ADH1 promoter driving expression of the GAL4-AD (*see* **Notes 1** and **2**).

5. cDNA library: a cDNA library fused to the GAL4-AD in pGAD424 (*see* **Note 3**).

6. 40% dextrose: weigh 200 g of dextrose, add distilled water to 500 mL, dissolve with mild heating (37°C), and filter sterilize. Store at room temperature.

7. YPD medium: 20 g of Bacto-peptone and 10 g of Bacto-yeast extract. Add distilled water to 950 mL, autoclave, and store at room temperature. Add 50 mL of 40% dextrose before use.

8. YPD-plates: 20 g of Bacto-peptone, 10 g of Bacto-yeast extract, and 17 g of agar. Add distilled water to 950 mL and autoclave. Allow cooling to 60°C, add 50 mL of 40% dextrose, and pour plates (about 40 mL per small plate; 10-cm diameter). Store plates upside-down in a plastic bag at 4°C.

9. 10× Dropout-TLH: 300 mg of L-isoleucine, 1500 mg of L-valine, 200 mg of L-adenine, 200 mg of L-arginine, 300 mg of L-lysine, 200 mg of L-methionine, 500 mg of L-phenylalanine, 300 mg of L-tyrosine, 200 mg of L-uracil, and 2000 mg of L-threonine. Dissolve in 1 L of distilled water and filter sterilize. Store at 4°C.

10. 100× His: 200 mg of L-histidine-HCl monohydrate. Dissolve in 100 mL of distilled water, autoclave, and store at 4°C.

11. 100× Leu: 1000 mg of L-leucine. Dissolve in 100 mL of distilled water and autoclave. Store at 4°C.

12. 100× Trp: 200 mg of L-tryptophan. Dissolve in 100 mL of distilled water and autoclave. Store at 4°C.

13. SD medium: 6.7 g of yeast nitrogen base without amino acids. Dissolve in 850 mL of distilled water and autoclave. Before use, add 100 mL of Dropout-TLH solution and 10 mL of other required amino acids. Add 50 mL of Dextrose. Store at room temperature.

14. 3-AT (Sigma Chemical Co., St. Louis, MO): 1 M 3-aminotriazole. Filter sterilize. Store in the dark at 4°C. Toxic!

15. SD plates: 6.7 g of yeast nitrogen base without amino acids and 17 g of agar. Dissolve in 850 mL of distilled water and autoclave. Allow cooling to ± 60°C, add 100 mL of dropout TLH and 10 mL of other required amino acids, and 50 mL of 40% dextrose. Add the required amount of 3-AT. Pour plates: about 50 mL per small plate (10 cm in diameter) or about 150 mL per large plate (14 cm in diameter). Store plates upside-down in a bag at 4°C.

16. Luria Broth (LB)–agar: 1% Bacto-Tryptone, 0.5% yeast extract, 0.5% NaCl, and 1.2% agar. Autoclave and allow cooling to 60°C. Prepare large plates containing LB–agar medium supplemented with the correct antibiotic.

17. LB medium: 1% Bacto-Tryptone, 0.5% yeast extract, and 0.5% NaCl. Autoclave, allow cooling, and supplement with the correct antibiotic.

2.2. Solutions

1. 10× TE: 100 mM Tris-HCl, pH 7.5; 10 mM ethylenediamine tetraacetic acid (EDTA) in distilled water. Autoclave and store at room temperature. For 1× TE, make fresh dilution in sterile distilled water.
2, 10× LiAc: 1 M LiAc, pH 7.5, with diluted acetic acid. Autoclave and store at room temperature.
3. TE/LiAc: Dilute 10× TE and 10× LiAc in sterile distilled water just before use.
4. 50% polyethyleneglycol (PEG): 50% (w/v) PEG4000 in distilled water. Autoclave and store at room temperature (*see* **Note 4**).
5. PEG/TE/LiAc: Combine 1 mL 10× TE, 1 mL 1× LiAc, and 8 mL 50% PEG just before use.
6. Z-buffer: 13.9 g $Na_2HPO_4 \cdot H_2O$, 750 mg KCl, 246 mg $MgSO_4 \cdot 7H_2O$, pH 7 with NaOH. Store at room temperature.
7. X-gal solution: 20 mg/mL in N,N-dimethylformamide; do not use plastic pipets or tubes. Store in the dark at –20°C. Toxic!
8. Complete Z-buffer: add 270 µL of β-mercaptoethanol and 1.67 mL of X-gal solution to 100 mL of Z-buffer. Make fresh when required. Keep in the dark. Toxic.
9. SCE buffer: 1 M sorbitol, 0.1 M Na_3 citrate, 10 mM EDTA, pH 6.5, to 7 in distilled water. Upon autoclaving, this buffer will turn slightly yellow. Store at room temperature.
10. SCEM buffer: 100 mL SCE buffer + 214 µL of β-mercaptoethanol. Prepare fresh when needed.
11. Yeast extraction buffer: 0.05 M Tris-HCl, pH 7.5, 0.025 M EDTA, 0.25 M NaCl, and 1% (w/v) sodium dodecyl sulfate (SDS) in distilled water. Autoclave and store at room temperature.
12. Sterile distilled water.
13. Dimethyl sulfoxide (DMSO).
14. Phenol:chloroform:isoamylalcohol (25:24:1): saturate with 10 mM Tris-HCl, pH 8.
15. 2× Laemmli loading buffer: 4% SDS, 125 mM Tris-HCl, pH 6.8, 20% glycerol, 10% β-mercaptoethanol, and 0.25% bromophenol blue.
16. 3 M NaAc.

2.3. Equipment and Other Items Required

1. 30°C incubator.
2. 30°C shaker.

3. Sterile glass beads.
4. Spectrophotometer equipped with 600-nm filter.
5. Microcentrifuge.
6. Sorvall centrifuge.
7. Water bath with adjustable temperature.
8. Thermomixer.
9. Speedvac.
10. Equipment for SDS-polyacrylamide gel electrophoresis and immuno-blotting.
11. Liquid nitrogen.
12. Nitrocellulose filters of the same size as the plates used.
13. Whatman filters of the same size as the plates used.

3. Methods

3.1. Preparative Steps

3.1.1. Construction of the GAL4-BD:Bait Gene Fusion Plasmid

To make the bait plasmid, clone the gene with which you would like to do a screening in frame with the GAL4-BD in the pAS2 vector (*see* **Note 5**). If you have one available, clone the gene of a known interacting protein (prey) in frame with the GAL4-AD in the pGAD424 vector.

3.1.2. Testing the GAL4-BD:Bait Gene Fusion Plasmid for Autoactivation and Expression

3.1.2.1. Test Transformation Using the LiAc Method

For a total of 30 transformations:

1. Grow yeast cells overnight at 30°C with vigorous shaking (250 rpm in a rotary shaker) in 20 mL of YPD medium.
2. In the morning, dilute the yeast cells in 300 mL of YPD to an OD_{600} value between 0.2 and 0.3. Allow them to grow at 30°C with vigorous shaking.
3. Monitor the growth of yeast cells by checking OD_{600} values at regular times. At an OD_{600} value between 0.6 and 0.8 (around 0.7 is optimal;

approx after 4 h), centrifuge the yeast cells in a Sorvall centrifuge at 1000g for 5 min.

4. Discard the supernatant and wash the cells twice in sterile distilled water, using 50 mL of water for each wash, taking care to resuspend the cells during each wash.

5. Centrifuge the cells at 1000g for 5 min and discard the supernatant carefully.

6. Resuspend the cells in 6 mL of TE/LiAc. Cells are now competent and remain competent when kept at room temperature for about 2 h.

7. Prepare the transformation mix in sterile Eppendorf tubes:

   0.1 μg of each of the appropriate plasmids

   10 μL of carrier DNA at 10 μg/μL (this carrier DNA should be denatured by keeping it at 95°C for 15 min, followed by rapid transfer to ice (*see* **Note 6**)

   Add 200 μL of the yeast cells prepared in **step 6**

   Add 600 μL of PEG/TE/LiAc

   Add 25 μL of DMSO

8. Mix the tubes by inverting and flicking the tubes.

9. Incubate the tubes at 30°C for 15 min, shake the tubes, and incubate for another 15 min at 30°C.

10. Transfer the tubes to a 42°C water bath for 15 min.

11. Cool the tubes on ice.

12. Centrifuge at 1000g for 15 min.

13. Remove the supernatant as much as possible and resuspend the cells in 300 μL 1× TE.

14. Plate the transformation mix on SD plates lacking the appropriate amino acids (*see* **Note 7**).

15. Incubate the plates inverted at 30°C for 2 to 4 d.

For testing your GAL4-BD:bait plasmid, transform the following plasmid combinations into yeast:

1: empty pAS2 vector   + empty pGAD424 vector
2: pAS2-bait vector   + empty pGAD424 vector
3: empty pAS2 vector   + pGAD424-prey vector
4: pAS2-bait vector   + pGAD424-prey vector

Plate equal amounts of the transformation mixtures on SD plates lacking Trp and Leu as well as on SD plates lacking Trp, Leu, and

His with or without 3-AT (*see* **Note 7**). Samples 1, 2, and 3 are controls for autoactivation and should only grow on SD plates lacking Trp and Leu. Sample 4 is a test for interaction and should also grow on plates lacking His.

3.1.2.2. EXPRESSION OF FUSION PROTEINS

1. Pick individual colonies of transformed yeast and grow them overnight in 10 mL SD medium lacking the appropriate amino acid (*see* **Note 8**). For analyzing the expression of the bait protein, use SD medium lacking Trp, for analyzing the expression of the interacting protein, use SD medium lacking Leu.
2. Centrifuge 6 mL yeast culture at 3000 rpm for 5 min.
3. Wash the pellet with distilled water.
4. Boil the pellet for 5 min in 200 µL 2× Laemmli loading buffer.
5. 50 µL of this sample is separated on a SDS-polyacrylamide gel and blotted onto a nitrocellulose filter. Perform immunoblotting with specific antibodies against your cloned genes or with antibodies specific for the GAL4-BD or the GAL4-AD (*see* **Note 9**).

3.1.2.3. BLUE STAINING

Only test colonies growing on SD plates lacking Leu, Trp, and His for blue staining (*see* **Note 10**).

1. A replica of yeast colonies grown on SD plates is made by gently laying a nitrocellulose filter on the plate. Do not move the filter once it lies on the plate. The filter will become wet.
2. Number empty plates according to the plates that are assayed, and put a Whatman filter paper inside. Wet the Whatman filter paper with 1.5 mL of freshly made complete Z-buffer.
3. Gently lift the nitrocellulose filter from the yeast plate and put it in liquid nitrogen for a couple of seconds, until the hissing of the nitrogen stops. Allow the filter to thaw, and then put it on top of the Whatman filter, yeast cells facing up.
4. Incubate the plates with filters in the dark at room temperature for at least 1 h, until colonies turn blue. Always include a positive and a negative control as a reference (*see* **Note 11**).
5. Stop the staining reaction by drying the filters in a fume hood.

3.1.3. Amplifying the GAL4-AD:cDNA Library

When amplifying a cDNA library, it is important to amplify all clones of the cDNA library. Therefore, it is safe to work with 10 to 20 times the amount of clones present in the library. Most GAL4-AD:cDNA libraries have complexities of $1-2 \times 10^6$ independent clones, so 10 to 20 $\times 10^6$ colonies should be amplified.

1. Determine how many single colonies can be plated on your large plates by plating dilutions of bacterial cultures. Petri dishes of 15 cm in diameter can bear approx 1×10^5 colonies, and plates of 22 cm $\times$ 22 cm can bear approx 3×10^5 colonies.
2. Transform the library plasmid DNA in a suitable strain of *Escherichia coli*. Determine the titer of the *E. coli* library transformants by plating 0.02, 0.2, and 2 µL of the culture onto small LB–agar plates. Keep the rest of the transformed culture at 4°C. Incubate the plates overnight upside-down at 37°C. Count the bacterial colonies on the titer plates and calculate the titer.
3. Plate the rest of the bacterial transformants at the maximal density of your plates (e.g., 1×10^5 for 15-cm plates). Incubate the plates upside-down at 37°C and allow the colonies to form (16–24 h).
4. Harvest the bacterial colonies by flooding one of the plates with 10 mL of sterile LB medium and scraping the colonies from the agar surface using a rubber policeman or bent glass rod. Be careful not to damage the agar while harvesting bacteria. Transfer the liquid to another plate and repeat. Each 10-mL aliquot can be used for up to five plates.
5. After the solution is saturated with bacteria, transfer to a centrifuge tube.
6. Start the next set of plates with a fresh 10 mL of LB medium. Repeat until all the plates have been scraped.
7. Mix all aliquots of bacteria together.
8. Harvest the cells by centrifugation at 10,000*g* for 10 min and proceed with plasmid DNA extraction (*see* **Note 12**).

3.2. Screening Procedure

3.2.1. Library Transformation

3.2.1.1. TRANSFORMATION OF THE BAIT PLASMID (*SEE* **NOTE 13**)

Transform 1 µg of bait plasmid to yeast (**Subheading 3.1.2.1.**) and plate the transformation mix on a SD plate lacking Trp. After 48 h, colonies will appear. Pick four or five individual transformed colonies and check expression of the bait protein as described in **Subheading 3.1.2.2.** Select the colony with the strongest expression for the cDNA library screening.

3.2.1.2. Transformation of the cDNA Library

1. In the morning, inoculate yeast cells already containing the bait plasmid to 20 mL of SD medium lacking Trp. Grow cells for 36–48 h at 30°C with shaking until the culture has an OD_{600} around 0.2. Do not allow the culture to become too dense because then the cells stop growing and they will take a lot of time to restart growing when diluted in **step 2**.
2. Next day in the evening, dilute the culture to about 200 mL of SD medium lacking Trp.
3. At the day of the library transformation, dilute the culture of bait containing yeast cells in 400 mL of YPD medium. OD_{600} should be around 0.2 in a total volume of 600 mL. Allow growing to an OD_{600} of about 0.7–0.75. This will take about 4 h.
4. Transfer cells into 4×250-mL bottles. Spin cells at 1000*g* for 5 min.
5. Remove supernatant and resuspend cells of each bottle in 50 mL of water. Spin cells at 1000*g* for 5 min.
6. Remove supernatant and resuspend cells of each bottle in 10 mL of freshly made 1× TE/LiAc. Transfer to a 50-mL tube. Spin at 1000*g* for 5 min.
7. Remove supernatant and bring final volume in each tube up to 2 mL with 1× TE/LiAc.
8. Do transformations in four 10× Eppendorf test tubes containing the following:
   200 µL of yeast cells
   10 µL of library DNA (*see* **Note 14**)
   30 µL of denatured carrier DNA (10 mg/mL; *see* **Note 6**)
   3.8 µL of 10× LiAc
   3.8 µL 10× TE
   30 µL of DMSO
   1.2 mL of PEG/TE/LiAc (freshly made)
9. Vortex and flick to assure DNA is well distributed.
10. Incubate at 30°C for 30 min while gently shaking in thermomixer, flick every 10 min.
11. Incubate at 42°C for 15 min.
12. Chill cells for 1 min on ice and spin at 1000*g* for 1 min.

13. Resuspend cells of each test tube in 300 µL of 1× TE.
14. Plate out an equal amount (approx 100 µL) from each test tube onto three plates lacking Leu, Trp, and His and containing the appropriate concentration of 3-AT (*see* **Note 15**).
15. Dilute 1 (1 of the transformation mix in 1× TE and plate out dilutions (1/10³, 1/10⁴, and 1/10⁵) on SD-plates lacking Trp and Leu to determine transformation efficiency (*see* **Note 14**).
16. Put plates at 30°C.

3.2.2. Picking Positive Colonies

Between 4 d and 3 wk after transformation of the GAL4-AD:cDNA library, a large number of very small colonies will appear on the plates. Mostly also a fewer number of larger colonies will appear (*see* **Note 16**).

1. Pick all these individual "larger" colonies and transfer them to individual wells of a 96-well plate containing 200 µL of SD medium lacking Trp and Leu. Incubate these 96-well plates for 2 h at 30°C. Spot a replica of the contents of the 96-well plates on a SD plate lacking Leu, Trp, and His, on which a sterile nitrocellulose filter has been placed. Spot also a replica on a backup SD plate lacking Trp, Leu and His. Incubate these plates at 30°C for approx 2 d or until clear yeast colonies appear. Mark the orientation of the nitrocellulose filters and lift them from the plates to perform a blue staining as described in **Subheading 3.1.2.3.**, starting at **step 2**.
2. The remaining "smaller" colonies are directly lifted from the original screening plates as described in **Subheading 3.1.2.3.** starting at **step 1**. Clearly indicate the orientation of the nitrocellulose filter and work under sterile conditions because these plates function as backup.

3.3. Characterization of Positives

3.3.1. Isolation and Analysis of GAL4-AD:cDNA Library Plasmid

3.3.1.1. RECOVERING PLASMID DNA FROM YEAST CELLS

1. Pick a positive colony from the backup plate and grow it for 1 d at 30°C in 5 to 10 mL of SD medium lacking Leu (*see* **Note 17**). Centrifuge at 1000*g* for 5 min.

2. Resuspend the cells in 5 mL of SCE buffer.
3. Centrifuge at 2800 rpm for 5 min.
4. Resuspend the cells in 1 mL of SCEM buffer to which some lyticase has been added and transfer to an Eppendorf tube.
5. Incubate for 1 h at 37°C.
6. Centrifuge for 5 min at 1500 rpm.
7. Remove supernatant carefully.
8. Resuspend pellet in 400 µL of yeast extraction buffer and homogenize carefully.
9. Incubate at 65°C for 15 min.
10. Extract with phenol:chloroform:isoamylalcohol until the interphase is clear, and then extract once with chloroform.
11. Precipitate DNA by adding 1/10 volume 3 *M* NaAc and 2.5 volumes ice-cold ethanol.
12. Centrifuge at 14000 rpm at 4°C for 15 to 30 min.
13. Wash the DNA by adding 750 µL of 70% ethanol and centrifuge at 14,000 rpm at 4°C for 15 min.
14. Dry the DNA for 10 min in a SpeedVac.
15. Dissolve the DNA in 100 µL of distilled water.

3.3.1.2. TRANSFORMATION OF BACTERIA AND INSERT ANALYSIS

Use 1 µL of plasmid DNA generated in the previous section to electrotransform the bacterial strain MC1061 or DH5α. Plate the bacteria on plates containing the required selection marker (Ampicilin for pGAD424; *see* **Notes 17** and **18**). Pick at least three colonies per plate, grow them overnight in LB medium, and isolate the plasmid. Check the length of the insert via restriction digest and agarose gel electrophoresis. Sequence the insert with primers directed against the sequence of the GAL4-AD (*see* **Note 19**). Check the length of the open reading frame and the nature of the insert (*see* **Note 20**).

3.3.2. Confirmation of the Interaction

Transform yeast cells according to **Subheading 3.1.2.1.** to check whether the interaction can be reproduced. Cotransform each isolated prey plasmid with your bait plasmid and include a negative control with the empty pAS2 plasmid. Plate transformation mixes on SD plates lacking Trp and Leu and SD plates lacking Trp, Leu, and His. Only candidates that do not grow on SD plates lacking Leu, Trp, and His when co-transformed with the empty pAS2 plasmid and do grow and stain blue on SD plates lacking Trp, Leu, and His when co-transformed with the bait plasmid should be further considered as real positives (*see* **Note 21**).

4. Notes

1. Before starting the yeast two-hybrid screening, the choice of the most appropriate vector system must be considered. A large number of GAL4-BD or GAL4-AD containing vectors exist. Probably the most important parameter is the protein expression level regulated by the alcohol dehydrogenase (ADH)1 promoter or variants thereof. The full-length ADH1 promoter, as found in the plasmids pAS1, pAS2, pAS2-1, pGADGH, pGAD1318, and pGADT7, leads to high-level expression. Other plasmids, including pGBT9, pGBKT7, pGAD424, pGAD10, pGAD-GL, and pACT2, contain a truncated ADH1 promoter that leads to low expression of the fusion protein. Lower expression levels might be important when proteins that are toxic for the yeast cells or that interfere with the endogenous yeast metabolism, have to be expressed.

2. The vectors for expression of GAL4-AD or GAL4-BD fusion proteins contain amino acids 768-881 or 1-147 of the GAL4 transcription factor, respectively.

3. It is best to start with a library from a tissue or cell type in which the bait protein is known to be expressed and to be biologically relevant. In our screening with the anti-apoptotic protein A20 as bait, we used an ad random primed and oligo-dT primed cDNA-library derived from mRNA of TNF-resistant fibrosarcoma L929R2 cells treated with TNF and cycloheximide during 4 h (*9*). Under these conditions, a clear induction of A20-mRNA can be observed in these cells. Many cDNA libraries cloned in frame of the GAL4-AD are commercially available.

4. The quality of PEG stock buffer decreases over time and it is advisable to prepare fresh stocks for library transformation.

5. Either full-length cDNA or only cDNA encoding the domain of interest of your protein can be cloned in frame with the GAL4-BD cDNA. In addition, because the fusion protein must translocate to the

nucleus, deletion of specific localization sequences present in your bait has to be considered.

6. A critical parameter for a good transformation is the quality of carrier DNA. This DNA must be of high molecular weight and completely denatured. You can shear DNA yourself and test which batch results in the best transformation efficiency. We used herring testes carrier DNA from Clontech. Although this DNA is already denatured, we repeat for just before each transformation the cycle of boiling for 15 min and chilling immediately on ice.

7. In many yeast strains *HIS3* has a leaky expression. The basal expression of this gene can be repressed by adding 3-aminotriazole (3-AT), an inhibitor of the *HIS3* gene product, to the medium. In case of autoactivation, 3-AT can reduce this problem by altering the sensitivity of detection. Therefore, in the test for autoactivation, yeast cells should be plated on medium lacking Trp, Leu, and His but containing increasing amounts of 3-AT (e.g., 5, 10, 25, and 50 m*M*). Optimization of the amount of 3-AT that is needed is necessary, because too-low concentrations will give many false-positives, whereas too-high concentrations will result in the loss of detection of weak interactions.

8. The ADH1 promoter used for the expression of the GAL4 fusion proteins normally drives the expression of the metabolic enzyme alcohol dehydrogenase 1. Expression from this promoter is optimal during the logarithmic growth of the yeast cells and becomes repressed during late log phase. For that reason it is best to test expression of the fusion proteins in logarithmic growing cells.

9. Some vectors, such as pAS2, contain the hemagglutinin (HA) tag, which allows immunological detection of the fusion protein with an anti-HA antibody. This HA tag is deleted in the vector pAS2-1. The same HA tag has also been introduced in pACT2, which is the successor of pGAD424. The MATCHMAKER System 3 (Clontech, Palo Alto, CA) fusion vectors, pGBKT7 and pGADT7, allow expression of fusion proteins with either c-Myc or HA epitope tags, so you can easily detect both fusion proteins.

10. When performing a β-galactosidase test, logarithmic growing yeast cells will give the best results. Therefore we lift yeast colonies of approx 1–2 mm in diameter. Yeast colonies that become pink by a lack of adenine in the medium stop growing and should not be used in the β-galactosidase test. The yeast cells are lifted on nitrocellulose filters (Bio-Rad, Hercules, CA, or Sartorius AG, Göttingen, Germany).

On these filters strong interactions normally become visible after 2 h, intermediate interactions after 6 h, whereas weak interactions might require up to 24 h of incubation. We observed that filters from other sources might need longer incubation times to reveal blue colonies.

11. If expression of the bait fusion protein results in autoactivation even at high concentrations of 3-AT, you can try to solve this problem by using only the important domain of your bait (*see* **Note 4**). Alternatively, you can try to solve the problem of auto-activation by cloning your bait behind a weaker promoter (e.g., a truncated ADH1 promoter; *see* **Note 1**).

12. There is no need to purify the plasmid from the endogenous RNA because it does not affect yeast transformation.

13. Co-transformation of library plasmid DNA with a GAL4-BD plasmid is not as efficient as transformation of the library DNA into a yeast strain already containing the GAL4-BD plasmid. Therefore, it is recommended that the GAL4-BD plasmid be transformed first into the yeast, followed by transformation of the library. In cases where the GAL4-BD fusion plasmid affects the growth of the yeast strain, it may be advisable to co-transform the GAL4-BD fusion plasmid and the library plasmid DNA.

14. During library transformation, high transformation efficiency is very important. The intention is to obtain as much transformants as possible but avoiding transformation of a single yeast cell with multiple library plasmids. Therefore, a library transformation efficiency test should be performed before starting a large-scale screening. In this test, increasing amounts of the cDNA library plasmid DNA must be transformed in the yeast cells already containing the bait plasmid. The transformed yeasts are plated and the number of transformants/ µg of DNA should be calculated. To screen a library till saturation, more than $5–10 \times 10^6$ yeast transformants need to be screened.

15. Because afterwards the yeast colonies have to be lifted, the surface of the agar plates on which they grow needs to be even. To avoid accidental damage of the surface of the agar plates, we used glass beads for spreading the transformed yeast cells.

16. It seems that yeast colonies maintained on medium containing 3-AT have a reduced viability. Therefore, we suggest to cryopreserve your positives as soon as possible.

17. If single yeast cells were initially transformed with more than one GAL4-AD:cDNA-library plasmid, they might lose the plasmid

encoding the interaction partner by growing them in medium lacking only Leu. This can be avoided by growing the yeast cells in medium lacking Leu, Trp, and His. This ensures that the bait plasmid and the cDNA library plasmid encoding the interacting fusion protein are maintained. Selection for the cDNA library plasmid can thereafter be conducted in bacteria transformed with the plasmid DNA recovered from yeast. One method makes use of the bacterial strain HB101 that is unable to grow on media lacking Leu. Only transformation of a cDNA library plasmid containing the *LEU2* gene can rescue these bacteria. However, HB101 bacteria grow very slowly and do not give high plasmid yields. Furthermore, DNA obtained from this strain is often difficult to sequence. Therefore, we normally transform plasmid DNA isolated from HB101 into MC1061 or DH5α bacteria and prepare plasmid DNA from these bacteria for sequencing. Alternatively, one can use the plasmids pGBKT7 and pGADT7 of the MATCHMAKER Two-Hybrid System 3. These have distinct bacterial selection markers, Amp^r and Kan^r respectively, which allows direct and specific selection for each plasmid in MC1061 or DH5α bacteria.

18. The plasmid DNA obtained from yeast is often not very clean, and difficulties in transforming bacteria may occur. Therefore, we recommend using electroporation for bacterial transformation.

19. If pGAD424 is used as library plasmid, the following primers can be used for sequencing: 5'-ACCACTACAATGGATGATG-3' for the 5' end, 5'-TAAAAGAAGGCAAAACGATG-3' for the 3' end.

20. A list of false-positives that have frequently been identified in two-hybrid screenings can be found on the Internet at the Golemis Lab Home Page (http://www.fccc.edu/research/labs/golemis/main_false.html).

21. Protein–protein interactions that are identified by the yeast two-hybrid system should always be confirmed by independent biochemical assays, such as co-immunoprecipitation or the GST pull-down assay. In this context, the Matchmaker System 3 GAL4-BD and GAL4-AD fusion vectors pGBKT7 and pGADT7 (Clontech, Palo Alto, CA) contain T7 promoters in front of the inserts, allowing in vitro transcription and translation. Furthermore, the bait and prey are expressed as fusions with either c-Myc or HA epitope tags, respectively, which eliminates the need to generate antibodies to new proteins for confirmation of their interaction.

Acknowledgments

This work was supported by the FWO, IUAP, BOF, and an EC-TMR grant. S.V.H. and K.H. are research assistant and post-doctoral researcher with the FWO-Vlaanderen, respectively.

References

1. Fields, S. and Song, O. (1989) A novel genetic system to detect protein–protein interactions. *Nature* **340,** 245–246.
2. Chinnaiyan, A. M., O'Rourke, K., Tewari, M., and Dixit, V. M. (1995) FADD, a novel death domain-containing protein, interacts with the death domain of Fas and initiates apoptosis. *Cell* **81,** 505–512.
3. Yang, X., Khosravi-Far, R., Chang, H.Y., and Baltimore, D. (1997) Daxx, a novel Fas-binding protein that activates JNK and apoptosis. *Cell* **89,** 1067–1076.
4. Wallach, D., Boldin, M. P., Kovalenko, A. V., Malinin, N. L., Mett, I. L., and Camonis, J. H. (1998) The yeast two-hybrid screening technique and its use in the study of protein–protein interactions in apoptosis. *Curr. Opin. Immunol.* **10,** 131–136
5. De Valck, D., Jin, D.Y., Heyninck, K., Van de Craen, M., Contreras, R., Fiers, W., et al. (1999) The zinc finger protein A20 interacts with a novel anti-apoptotic protein which is cleaved by specific caspases. *Oncogene* **18,** 4182–4190.
6. Heyninck, K., De Valck, D., Vanden Berghe, W., Van Criekinge, W., Contreras, R., Fiers, W., et al. (1999) The zinc finger protein A20 inhibits TNF-induced NF-kappaB-dependent gene expression by interfering with an RIP- or TRAF2-mediated transactivation signal and directly binds to a novel NF-kappaB-inhibiting protein ABIN. *J. Cell Biol.* **145,** 1471–1482.
7. Van Huffel, S., Delaei, F., Heyninck, K., De Valck, D., and Beyaert, R. (2001) Identification of a novel A20-binding inhibitor of nuclear factor-κB activation termed ABIN-2. *J. Biol. Chem.* **276,** 30,216–30,223.
8. Beyaert, R., Heyninck, K., and Van Huffel, S. (2000) A20 and A20-binding proteins as cellular inhibitors of nuclear factor-kappaB-dependent gene expression and apoptosis. *Biochem. Pharmacol.* **60,** 1143–1151.
9. De Valck, D., Heyninck, K., Van Criekinge, W., Contreras, R., Beyaert, R., and Fiers, W. (1996) A20, an inhibitor of cell death, self-associates by its zinc finger domain. *FEBS Lett.* **384,** 61–64.

17

The Yeast Three-Hybrid System
As a Tool to Study Caspases

Wim van Criekinge, Peter Schotte,
Karen Heyninck, and Rudi Beyaert

Summary

Caspases are cysteine proteases that play an essential role during apoptotic cell death and inflammation. They are synthesized as catalytically dormant proenzymes, containing an N-terminal prodomain, a large subunit (p20) containing the active site cysteine, and a small subunit (p10). The active enzymes function as tetramers, consisting of two p20/p10 subunit heterodimers. Both subunits contribute residues that are essential for substrate recognition. Activation of caspases culminates in the cleavage of a set of cellular proteins, resulting in disassembly of the cell or proinflammatory cytokine production. Inappropriate caspase activation contributes to or accounts for several diseases. The identification of caspase-interacting proteins that might act as activators, substrates, or inhibitors is therefore an attractive step in the development of novel therapeutics. However, caspase substrates and other proteins that bind specifically with the active heterodimeric p20/p10 form of caspases will escape detection in a classical two-hybrid approach with an unprocessed caspase precursor as bait. Alternatively, a number of so-called three-hybrid systems to analyze more complex macromolecular interactions have been developed. We describe the use of a three-hybrid approach adapted to the needs of caspases to detect and analyze the interaction of mature heteromeric caspases with protein substrates or inhibitors.

Key Words: Three-hybrid; caspases; apoptosis; caspase-1; protein–protein interactions; IL-1.

From: *Methods in Molecular Biology, vol. 282: Apoptosis Methods and Protocols*
Edited by: H. J. M. Brady © Humana Press Inc., Totowa, NJ

1. Introduction

Caspases play an essential role during apoptotic cell death *(1)*. These enzymes define a new class of cysteine proteases and comprise a multigene family with more than a dozen distinct mammalian family members. Caspases are synthesized as catalytically dormant tripartite proenzymes. Each zymogen contains an N-terminal prodomain of variable length, a large subunit (approx 20 kDa = p20) containing the active site cysteine, and a C-terminal small subunit (approx 10 kDa = p10). An aspartate cleavage site separates the prodomain from the large subunit, and an interdomain linker containing one or two aspartate cleavage sites separates the large and small subunits (**Fig. 1**). Activation accompanies proteolysis of the interdomain linker and usually results in subsequent removal of the prodomain. The active enzymes function as tetramers, consisting of two large/small subunit heterodimers *(2,3)*. The heterodimers each contain an active site composed of residues from both the small and large subunits. Similarly, both subunits contribute residues that are essential for substrate recognition. Activation of caspases culminates in the cleavage of a set of cellular proteins, resulting in disassembly of the cell or proinflammatory cytokine production. Inappropriate caspase activation and apoptosis contributes to or accounts for several disease pathogeneses. In some cases, such as Alzheimer's disease, caspases also appear to play a role in pathogenic exacerbation by cleavage of disease-associated substrates. Several macromolecular caspase inhibitors have been identified, including the cowpox virus serpin CrmA *(4)*, and members of the inhibitor of apoptosis superfamily *(5)*. The identification of caspase-interacting proteins that might act as activators, substrates, or inhibitors is therefore an attractive step in the development of novel therapeutics.

The yeast two-hybrid system is a well-established method to identify and study genes encoding proteins that interact with a protein of interest *(6)*. Therefore, a caspase substrate or another protein that binds specifically with the active heterodimeric p20/p10 form of caspases will escape detection in a two-hybrid approach with an unprocessed caspase precursor as bait. A number of so-called three-

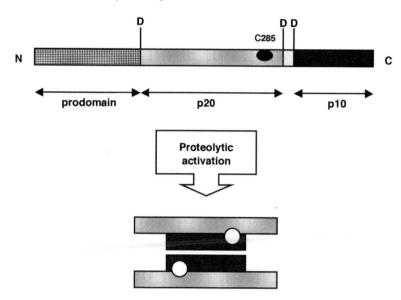

Fig. 1. General structure of procaspases and proteolytic maturation. Caspases are synthesized as single-chain precursors that undergo proteolytic maturation/activation by cleavage after specific Asp residues. This generates a prodomain, a large subunit (p20) and a small subunit (p10). The active enzyme is composed of a p10 and a p20 subunit, each of which contributes amino acids to the active site (shown by open circles). Two heterodimers associate to form a tetramer.

hybrid systems to analyze more complex macromolecular interactions have been developed *(7)*. We describe the use of a three-hybrid approach adapted to the needs of caspases to detect and analyze the interaction of mature heteromeric caspases with protein substrates or inhibitors. The usefulness of such a three-hybrid approach has been demonstrated with the specific interaction of heteromeric (p20/ p10) caspase-1 with the viral pseudosubstrate caspase inhibitors CrmA and p35, as well as with the prototype caspase-1 substrate prointerleukin-1β *(8)*. A similar approach can be used to screen a complete complimentary DNA (cDNA) library for proteins interacting with heteromeric caspases.

Because the materials and methods used in a three-hybrid experiment are basically identical to those that are used in a two-hybrid screening, these will not be described in detail here. The reader is referred to the accompanying chapter on yeast two-hybrid screening. In the present chapter, we will simply give an outline of the method and some useful advice that we have found to be important in our experiments. The method is based on four points:

1. Preparation of bait and prey expression plasmids.
2. Verification of the reconstitution of the heteromeric caspase in yeast.
3. Actual three-hybrid experiment and identification of positive clones.
4. Quantification of β-galactosidase activity.

2. Materials

1. Yeast strain HF7c.
2. Yeast plasmid pAS2 (Clontech, Palo Alto, CA), containing *TRP1* as selection marker, and the full-length ADH1 promoter driving expression of the GAL4 DNA-binding domain (GAL4-BD; amino acids 1–147).
3. Yeast plasmid pAS3 containing *ADE2* as selection marker and the full-length ADH1 promoter driving expression of the GAL4-BD (*see* **Note 1**). pAS3 was created by blunt ligation of an *ade2* gene under its own promoter into an *Eco*RV opened pAS2 plasmid. As a result, the *trp* gene of pAS2 is disrupted and the gene conferring cycloheximide sensitivity is removed (*see* **Notes 2** and **3**).
4. Yeast plasmid pGAD424 (Clontech, Palo Alto, CA), containing *LEU2* as selection marker, and the truncated ADH1 promoter driving expression of the GAL4 activation domain (GAL4-AD; amino acids 768–881).
5. β-Gal lysis buffer: 100 m*M* HEPES, 150 m*M* NaCl, 4.5 m*M* L-aspartic acid hemimagnesium salt, 1% bovine serum albumin, 0.1% Tween-20 (v/v) at pH 7.25. Filter sterilize.
6. β-Galactosidase lysis buffer with chlorophenolred-β-D-galactopyranoside (CPRG): dissolve 27.1 mg of CPRG in 20 mL of β-galactosidase lysis buffer (final concentration CPRG is 2.23 m*M*). Filter sterilize. Store in the dark at 4°C.
7. Growth media and other solutions or reagents, as well as equipment are identical to those used in a two-hybrid experiment (*see* Chapter 16).

3. Methods

3.1. Construction of the Bait and Prey Expression Plasmids

To detect or analyze proteins that bind to mature heteromeric caspase-1, both subunits (p20 and p10) have to be cloned in frame with the GAL4-BD in the pAS2 and pAS3 vector, respectively (*see* **Note 4**). The *CrmA* gene or another insert encoding a potential caspase-1 interacting protein is cloned in frame of the GAL4-AD in pGAD424. Cloning of p20, p10, and CrmA in frame of the GAL4-BD or GAL4-AD is achieved by introducing an additional *Nco*I and *Sal*I restriction site by polymerase chain reaction at their N- and C-terminal end, respectively.

In all experiments, an inactive caspase-1 mutant in which Cys^{285} in the catalytic site is mutated to Ser should be used (*see* **Note 5**). Several kits for site-directed mutagenesis are commercially available. Cloning and mutations should be verified by DNA sequencing.

3.2. Testing the Bait Proteins for Autoactivation, Expression, and Reconstitution of a Heteromeric Bait

In general, it is necessary to test the suitability of the constructs in a number of assays. The first one is to check that the bait is unable *per se* to activate transcription. Indeed, some constructs may prove to self-activate. The second assay is to assess whether the bait is properly expressed. A third assay is performed to verify in a yeast two-hybrid experiment whether p20 and p10 can indeed form a heteromeric complex in yeast.

3.2.1. Autoactivation

1. Transform competent HF7c yeast cells with both p20 and p10 bait plasmids and empty pGAD424.
2. Plate equal amounts of the transformation mixtures onto Leu⁻, Trp⁻, and Ade⁻ plates as well as onto Leu⁻, Trp⁻, Ade⁻, and His⁻ plates and incubate for 2–3 d at 30°C.
3. If colonies grow on Leu⁻, Trp⁻, Ade⁻, and His⁻ plates, perform a β-galactosidase assay on a replica of yeast colonies that is made on a nitrocellulose filter.

If yeast colonies do not grow on Leu⁻, Trp⁻, Ade⁻, and His⁻ plates, it means that the bait does not activate transcription *per se* (*see* **Note 6**).

3.2.2. Protein Expression

Lysates are prepared from cultures that were grown from individual colonies of transformed yeast, and analyzed for expression of the bait by sodium dodecyl sulfate-polyacrylamide gel electrophoresis and immunoblotting with antibodies raised against the GAL4-BD (available from several commercial sources).

3.2.3. Testing the Reconstitution of a Heteromeric Caspase

Before starting a three-hybrid experiment with p20/p10 caspase-1, it is advisable to test in a classical two-hybrid experiment whether the p20 and p10 subunits can form a heteromeric complex in yeast. The p20 and p10 subunits are therefore cloned in pAS2 and pGAD424, respectively, and transformed to HF7c. The ability of transformed cells to grow on His- plates and the expression of β-galactosidase activity is indicative for the formation of a heteromeric interaction between the p20 and p10 subunit of caspase-1. Blue colonies should be detectable within 5 to 7 h.

3.3. Testing a Three-Hybrid Interaction

The use of the three-hybrid system is exemplified with the interaction of mature caspase-1 with its inhibitor CrmA (**Fig. 2**). Several negative controls should be included.

1. Transform yeast cells with the following plasmid combinations to check whether interaction occurs:

    1: empty pAS2 + empty pAS3 + empty pGAD424
    2: pAS2-p20 + empty pAS3 + empty pGAD424
    3: empty pAS2 + pAS3-p10 + empty pGAD424
    4: pAS2-p20 + pAS3-p10 + empty pGAD424
    5: empty pAS2 + empty pAS3 + pGAD424-CrmA
    6: pAS2-p20 + empty pAS3 + pGAD424-CrmA

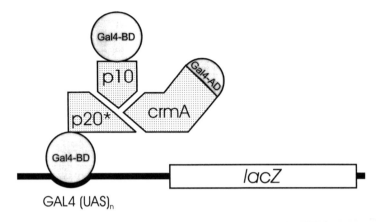

Fig. 2. Principle that forms the basis of a three-hybrid system with heteromeric baits. Here the bait consists of a heterodimer between fusion proteins of the GAL4-BD with the p20 and p10 subunit of caspase-1, respectively. Interaction of the pray (CrmA fused to the GAL4-AD) with caspase-1 requires both subunits. An asterisk in p20 indicates the Cys285Ser mutation that is introduced in order to inactivate the caspase.

      7: empty pAS2 + pAS3-p10 + pGAD424-CrmA
      8: pAS2-p20 + pAS3-p10 + pGAD424-CrmA

2. Plate equal amounts of the transformation mixtures onto Leu⁻, Trp⁻, and Ade⁻ plates, as well as onto Leu⁻, Trp⁻, Ade⁻, and His⁻ plates, and incubate for 2–3 d at 30°C.

3. Test colonies growing on Leu⁻, Trp⁻, Ade⁻, and His⁻ plates for blue staining by making a replica of the colonies on a nitrocellulose filter. After gently lifting the nitrocellulose filter from the yeast plate, put it in liquid nitrogen for 10 s.

4. Allow the filter to thaw at room temperature (2 min), and then put it in an empty plate on top of a Whatmann filter, which should be prewetted with 1.5 mL of freshly made complete Z buffer. Yeast cells should be facing up!

5. Incubate the filters in the dark at room temperature and check for the appearance of blue colonies at regular times. In the case of interaction, blue staining should be detectable within less than 18 h.

6. Stop the staining reaction by drying the filters in a fume hood (*see* **Note 7**).

3.4. Quantitative Colorimetric Assay
for β-Galactosidase Reporter Activity

Positive yeast colonies can be tested for the levels of lacZ expression using a liquid CPRG assay. This assay is based on the ability of the β-galactosidase reporter gene product to convert CPRG to chlorophenolred and D-galactose. This is associated with a change in color from yellow to red. The velocity of the reaction as well as the intensity of the red color are a measure for the amount of β-galactosidase that is present in yeast cells. Because this amount is dependent on the strength of the yeast two-hybrid interaction, the CPRG assay can be used to quantitate the strength of the interaction between two proteins in a yeast two- or three-hybrid assay.

1. Inoculate individual colonies (1–3 in mm diameter) from Leu⁻, Trp⁻, Ade⁻, and His⁻ plates into 5 mL of liquid Leu⁻, Trp⁻, Ade⁻, and His⁻ medium and incubate at 30°C overnight with vigorous shaking (230–250 rpm).
2. Inoculate the next morning 2 mL of this culture in 8 mL of fresh YPD medium and incubate further at 30°C for 3–5 h (230–250 rpm) until the OD_{600} is between 0.5 and 0.8 (cells in mid-log phase).
3. Mix the culture vigorously during 1 min to eliminate all cell clumps that may be present and determine the exact OD_{600}.
4. Pellet 1.5 mL of the culture in an Eppendorf tube (14,000 rpm, 30 s), remove the supernatant carefully, and resuspend the cell pellet in 1 mL of β-galactosidase lysis buffer.
5. Pellet again in an Eppendorf centrifuge (14,000 rpm, 30 s), remove the supernatant, and resuspend the pellet in 100 μL of β-galactosidase lysis buffer.
6. Snap freeze by placing the tubes in liquid nitrogen.
7. Thaw the tubes by incubation in a 37°C water bath.
8. Repeat the freeze\thaw cycle five times (*see* **Note 8**).
9. Add 900 μL of β-galactosidase lysis buffer containing 2.23 mM CPRG and mix strongly by vortexing (*see* **Notes 9** and **10**). Write down the exact time.
10. Incubate at room temperature until an orange–red color starts to develop (*see* **Note 11**).
11. Stop the reaction by the addition of 500 μL of 3 mM $ZnCl_2$ and record exactly the elapsed time.

12. Pellet the debris by centrifugation at 14,000 rpm for 1 min in an Eppendorf centrifuge. Carefully transfer the supernatant in a cuvette and determine the OD_{578}. Use a sample that did not contain any yeast lysate as a blank.

13. Calculate the units of β-galactosidase activity using the following formula (*see* **Note 12**):

$$Units = (1000 \times OD_{578})/(t \times V \times OD_{600})$$

where t is the elapsed time in minutes, V is the volume of culture used in mL, and OD_{600} is the optical density of a 1-mL culture.

4. Notes

1. Both baits (GAL4-BD:p20 and GAL4-BD:p10, respectively) should be under control of the same promoter or promoters of equal strength. This will result in equal amounts of both fusion proteins and maximal reconstitution of the heteromeric caspase. pAS2 and pAS3 are both regulated by the full-length alcohol dehydrogenase (ADH) 1 promoter, leading to high expression levels.

2. *CYH2* encodes the L29 protein of the yeast ribosome. Cycloheximide, a drug that blocks polypeptide elongation during translation, prevents the growth of cells that contain the wild-type *CYH2* gene. When a wild-type *CYH2* containing plasmid is present in a mutant yeast strain resistant to cycloheximide, it confers cycloheximide sensitivity. Adding cycloheximide to the medium results in a selection for yeast that has lost the *CYH2*-containing plasmid. This technique is sometimes used to select for only the library insert containing plasmid after screening. Because a plethora of alternatives is available (e.g., transformation of the isolated plasmid to bacteria and selection in Leu⁻ medium; *see* Chapter 16), and the leakage of the *CYH2* promoter is believed to be partially responsible for yeast growth retardation, this gene was removed from the pAS3 plasmid.

3. The disruption of the *Trp* gene is needed to allow selection for either pAS2 or pAS3. The use of pAS3 is fully compatible with other plasmids that are routinely used in the GAL4 two-hybrid system.

4. The cloning of p20, as well as p10, as a fusion protein with the GAL4-BD has a number of advantages in comparison to the use of only a single GAL4-BD fusion protein. The nuclear localization sequence that is present in the GAL4-BD ensures that both proteins will be properly transported into the nucleus, which is a prerequisite

for reporter gene activation. Moreover, the use of two GAL4-BD fusion proteins allows one to isolate proteins that also would specifically bind with p20 or p10 subunits, thus combining two-hybrid and three-hybrid screenings in a single experiment. Finally, it also eliminates constraints on the level of promotor binding. If problems are encountered because of the use of two fusion proteins (e.g., steric hindrance), one can try to express one of the caspase subunits without a GAL4-BD.

5. Coexpression of wild-type GAL4-BD:p20 and GAL4-BD:p10 results in very low transformation efficiencies and severe growth retardation as a result of the toxicity of active caspase-1 in yeast. Moreover, we have observed that coexpression of GAL4-BD:p20 and GAL4-BD:p10 in yeast leads to autoproteolytic cleavage of GAL4-BD in its C-terminal part, resulting in the release of the fused protein *(9)*.

6. Also included in this experiment are a positive and negative control for autoactivation. The positive control will grow and turn blue on His⁻ plates, whereas the negative control will not. These plasmids can be obtained either commercially (e.g., p53 and SV40-LT) or by contacting our laboratory. Because the His⁻ test is more sensitive than the β-galactosidase test, and leaky expression of HIS3 occurs in many yeast strains, it can be commonly observed that transformed yeasts do not turn blue, but grow on His⁻. Because the His⁻ test is the first to be used in a screening, one has to decide whether the generated background is too high or not to continue. Alternatively, yeast strains that are less responsive can be used, or the basal expression of the *His3* gene can be repressed by adding optimized amounts of 3-aminotriazole (5–50 mM), an inhibitor of the HIS3 gene product, to the medium.

7. All plasmid mixtures should give rise to colonies that grow on Leu⁻, Trp⁻, and Ade⁻, plates, which only serves as a control for transformation efficiency. If the prey (in this case CrmA) specifically interacts with the p20/p10 heteromer, only yeast cells transformed with plasmid mixture 8 should be able to form colonies on Leu⁻, Trp⁻, Ade⁻, and His⁻ plates. If colonies on His⁻ are also formed by yeast cells transformed with plasmid mixture 6 or 7, this would indicate that the prey is able to interact already with one of the subunits (which is not so in the case of CrmA).

8. Although other protocols mention that two cycles of freeze-thawing should be sufficient to open the cells, we have noted that five cycles of freeze thawing are mostly required to open all cells.

9. Add 1 mL of β-galactosidase lysis buffer to an empty Eppendorf tube, which will serve as a blank.

10. The reaction is started by the addition of CPRG. It is therefore crucial that the buffer with CPRG is added quickly and at the same time to all samples, and that the time of CPRG addition to the extract is exactly noted. It may therefore be practical to add the CPRG solution in precisely time intervals, and to stop the reaction with exactly the same time interval as before so that the incubation times are identical for all samples.

11. Follow the reaction carefully. In the case of strong interactions (e.g., p53 and SV40-LT), a change of color will already occur within seconds. Weak interactions might require several hours. Do not wait to stop the reaction until a fully red color has developed, as this will result in optical densities that are out of the linear range (linear range at $OD_{578} = 0.25–1.8$).

12. One unit of β-galactosidase is defined as the amount of β-galactosidase that is able to hydrolyze, in 1 min, 1 μmol of CPRG into chlorophenol red and D-galactose.

Acknowledgments

The project was funded by the FWO, IUAP, BOF, IWT, and an EC-TMR.

References

1. Wolf, B. B. and Green, D.R. (1999) Suicidal Tendencies: apoptotic cell death by caspase family proteinases. *J. Biol. Chem.* **274**, 20,049–20,052.

2. Wilson, K. P., Black, J.-A. F., Thomson, J. A., Kim, E. E., Griffith, J. P., Navia, M. A., et al. (1994). Structure and mechanism of interleukin-1β converting enzyme. *Nature* **370**, 270–275.

3. Walker, N. P., Talanian, R. V., Brady, K. D., Dang, L. C., Bump, N. J., Ferenz, C. R., et al. (1994). Crystal structure of the cysteine protease interleukin-1β _converting enzyme: a (p20/p10)2 homodimer. *Cell* **78**, 343–352.

4. Zhou, Q., Snipas, S., Orth, K., Muzio, M., Dixit, V. M., and Salvesen, G. S. (1997). Target protease specificity of the viral serpin CrmA. Analysis of five caspases. *J. Biol. Chem.* **272**, 7797–8590.

5. Deveraux, Q. L. and Reed, J .C. (1999) IAP family proteins—suppressors of apoptosis. *Gen. Dev.* **13,** 239–252.

6. Fields, S. and Song, O. (1989) A novel genetic system to detect protein-protein interactions. *Nature* **340,** 245–246.

7. Tirode, F., Malaguti, C., Romero, F., Attar, R., Camonis, J., and Egly, J. M. (1997). A conditionally expressed third partner stabilizes or prevents the formation of a transcriptional activator in a three-hybrid system. *J. Biol. Chem.* **272,** 22,995–22,999.

8. Van Criekinge, W., Van Gurp, M., Decoster, E., Schotte, P., Van de Craen, M., Vandenabeele, P., et al. (1998) Use of the yeast three-hybrid system as a new tool to study caspases. *Anal. Biochem.* **263,** 62–66.

9. Van Criekinge, W., Cornelis, S., Van de Craen, M., Vandenabeele, P., Fiers, W., and Beyaert, R. (1999) GAL4 is a substrate for caspases: implications for two-hybrid screening and other GAL4-based assays. *Mol. Cell Biol. Res. Commun.* **1,** 158–161.

18

Cloning of Apoptosis-Related Genes by Representational Difference Analysis of cDNA

Michael Hubank, Fredrik Bryntesson, Jennifer Regan, and David G. Schatz

Summary

Apoptosis is frequently triggered by events that alter the expression of key target genes. Under these circumstances, the genes involved can be identified by techniques that analyze gene expression. Researchers now have a choice of reliable and effective methods for differential gene expression analysis. Comparative approaches, including gene microarray analysis, serial analysis of gene expression, and differential display provide global information about expression levels. Subtractive approaches like complementary DNA representational difference analysis (cDNA RDA) and suppression subtraction polymerase chain reaction identify a focused set of differentially expressed genes. The most suitable technique to apply depends on individual circumstances. cDNA RDA is particularly useful in nonstandard model organisms for which comprehensive gene microarrays are not available and is best used for the identification of genes with a large difference in expression levels between two populations. The technique involves the generation of amplified mixtures of cDNA fragments that are typically smaller than 1000 base pairs and represent >86% of mRNA species from each starting population. Transcriptional differences between two populations can then be identified by subtraction of cDNA amplicons followed by further polymerase chain reaction amplification. The technique is capable of detecting differences for genes expressed at less than one copy per cell and is achievable using standard laboratory apparatus. cDNA RDA can identify genes not previously described in the database, can detect low abundance transcripts (e.g., from mixed cell populations), and is best applied in experiments where relatively few differentially expressed genes are expected. Here, we describe the application of cDNA RDA to the identification of apoptosis-related genes.

From: *Methods in Molecular Biology, vol. 282: Apoptosis Methods and Protocols*
Edited by: H. J. M. Brady © Humana Press Inc., Totowa, NJ

Key Words: Representational difference analysis; gene expression profiling; apoptosis; subtractive hybridization; differential gene expression.

1. Introduction

In metazoan cells, the opposing activities of death-inducing and survival-promoting proteins establish an equilibrium that determines cell survival. Apoptosis is triggered when this balance is altered in favor of death promoters like Bax and Bim, and against survival genes such as *Bcl-2 (1)*. In many systems, events that precipitate apoptosis affect a change in the levels of key proteins by activating signaling pathways that culminate in altered gene expression. Changes in gene regulation can occur in response to death-promoting signals, as is the case in the p53-mediated DNA damage response *(2)* or result from the lack of survival signals, which occurs during immune development *(3)*. Because these events are controlled at the transcriptional level, the genes involved can be identified by techniques that analyze gene expression.

Researchers now have a choice of several reliable and effective methods for differential gene expression analysis. Many of these techniques are now well established and include comparative approaches, such as microarray analysis, serial analysis of gene expression, and differential display, and subtractive approaches like complementary DNA representational difference analysis (cDNA RDA), suppression subtraction polymerase chain reaction (PCR), and cDNA or RNA library subtraction. The most suitable technique to apply depends on individual circumstances, and careful consideration should always be given to the limitations and capabilities of both the experimental system and the intended method of analysis.

This chapter describes the use of cDNA RDA to identify apoptosis-related genes *(4,5)*. cDNA RDA has a successful track record of gene discovery *(6–9)* and has been applied successfully by several groups to discover apoptosis-related genes *(10–12)*. cDNA RDA involves the generation of amplified mixtures of cDNA fragments that are typically smaller than 1000 base pairs and represent >86% of mRNA species from each starting population (**Fig. 1**). Transcriptional differences between two starting populations can then be

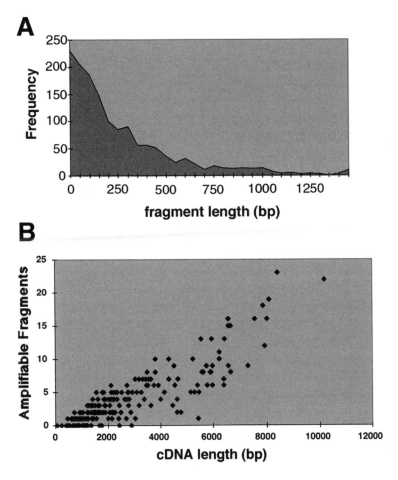

Fig. 1. Analysis of 200 randomly selected full-length mammalian cDNA sequences for the presence of potentially amplifiable *Dpn*II fragments. (**A**) Length of *Dpn*II fragments: the majority of fragments are below 600 base pairs in length. Amplifiable fragments fall into the range of approx 110 to 1200 base pairs. (**B**) Numbers of amplifiable fragments as a function of cDNA size: 86% of genes contain at least one amplifiable fragment.

identified by subtraction of the amplified cDNA populations followed by amplification by PCR. The technique is capable of detecting differences of genes expressed at less than one copy per cell and is achievable using standard laboratory apparatus.

cDNA RDA is ideal for apoptosis-related experiments where the starting quantities of RNA are often low. cDNA RDA is best used for the identification of genes that are switched on in one population but off in another, to identify genes not previously described in the database, and to identify low abundance transcripts. It should be applied to experiments where relatively low numbers of genes are expected to be differentially expressed. In situations where comprehensive information on gene transcription is desired, or differences in expression of less than fivefold are the target, microarray analysis would be the preferred approach.

Application of cDNA RDA to detect apoptosis-related genes requires some special considerations. We will outline appropriate experimental design, and describe protocols for the generation of representative PCR amplicons from apoptosis-positive and -negative populations, the subtractive hybridization of the representative amplicons, and the cloning and screening of the resulting products (**Fig. 2**).

1.1. Experimental Design

The ideal cDNA RDA experiment consists of a well-defined, precisely controlled biological system in which the expression of a targeted subset of genes is induced or repressed on a given signal. cDNA RDA is an extremely sensitive technique and often leads to the detection of genes with very low levels of expression. Care should therefore be taken to minimize secondary effects, nonspecific transcriptional differences, and heterogeneous differences that arise from sources other than the apoptotic signal. Subtractions should preferably be performed on cell lines separated in their origin by the shortest time and the fewest possible manipulations, and differences should be screened against several independent isolates to rule out random changes. Where tissue samples form the basis of the subtraction, dissection must be accurate, and the subjects should be as closely related as possible (e.g., littermates). It is also important to give consideration to the methods of secondary screening that will be used. Secondary screening involves the functional investigation of genes after they have been established as genuine and reproducible differences by primary screening. Establishment

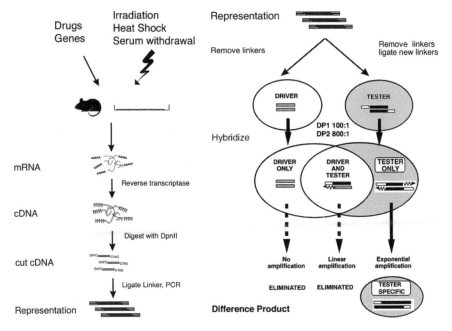

Fig. 2. cDNA RDA scheme.

of a functional analysis system in advance can often lead to better-targeted subtraction experiments. Certain approaches often applied in apoptosis-related experiments require special consideration.

1.1.1. Fluorescence-Activated Cell Sorting (FACS)

FACS populations of cells form ideal substrates for cDNA RDA (*see* Chapter 3), but care is required to keep cross contamination of populations to a minimum. Sorting should be performed on slow sort rate settings that isolate highly pure populations because quality is more important than yield. Sort gates should be set as far apart as possible, and re-sorts should be performed to assess purity. It is frequently possible to enrich cells by magnetic beads approaches before FACS to reduce sort times, but where long sort times are required, the cells should be kept on ice and sorted into chilled medium. The numbers of cells required depends on their size and activity, but typically 1–10 μg of total RNA will be sufficient.

This corresponds to approx 1×10^6 resting lymphocytes. Smaller samples can be used, but care must be taken to verify the representative nature of the resulting amplicons.

1.1.2. Induced Apoptosis

Apoptosis is frequently induced in experimental systems by a defined signal, such as ionizing radiation, dexamethasone, or camptothecin. In these cases, it is important to differentiate between first-round transcriptional targets of the inducing agents and subsequent "knock-on" effects. This can be accomplished by inclusion of a protein synthesis inhibitor (cycloheximide), but it is important to generate representations from both cycloheximide-treated and untreated controls to ensure that identified transcripts are not artefacts of treatment with the inhibitor. In some systems, apoptosis can result from the expression of an exogenous transgene, which can either be inducible or show constitutive expression. Appropriate controls include the preparation of representations from vector-only transfectants, with or without inducing agents. Control representations can then be mixed in the Driver to prevent the cloning of unwanted transcripts such as antibiotic resistance marker (e.g., *Neo*).

2. Materials

2.1. Instrumentation and General Chemicals

1. Equipment: Refrigerated microcentrifuge; PCR machine with heated lid; Ligation water bath; Polytron homogenizer.
2. Chemicals: Chloroform; isopropanol; glycogen (Roche); Phenol: chloroform:isoamyl alcohol (P:C:I; 25:24:1) (hazardous); chloroform: isoamyl alcohol (C:I; 24:1); 100%; and 70% ethanol.

2.2. RNA Isolation

1. Chemicals: Trizol reagent (Life Technologies Ltd; hazardous); RNAse-free water.
2. Kit: Oligotex mRNA isolation kit, which includes 2× binding buffer, oligotex suspension, wash, and elution buffers (Qiagen Ltd).

2.3. cDNA Synthesis

1. Removal of genomic DNA: 0.1 *M* dithiothreitol (DTT); 3 *M* sodium acetate, pH 5.3; 50 m*M* MgCl$_2$; RNase Inhibitor (Promega; 20-40 U/μL); DNase I (RNase-free; Roche; 10-50 U/μL).
2. First strand synthesis: 5× first-strand buffer (e.g., Superscript II buffer, Life Technologies); dNTP mixture (20 m*M* each of dATP, dCTP, dGTP, and TTP), OligodT primer (Promega; 50 ng/μL), RNase inhibitor (20–40 U/μL), 15 m*M* βNAD (Roche); and reverse transcriptase (RT, e.g., Superscript II, Life Technologies, 200 U/μL).
3. Second-strand synthesis: 5× RT2 second-strand buffer (1 mL of 5× RT2 is prepared by mixing 100 μL of 1 *M* Tris-HCl (pH 7.5), 500 μL of 1 *M* KCl, 25 μL of 1 *M* MgCl$_2$, 50 μL of 1 *M* (NH$_4$)$_2$SO$_4$, 50 μL of 1 *M* DTT, 50 μL of 5 mg/mL bovine serum albumin, and 225 μL of water); *Escherichia coli* DNA ligase (NEB, 5 U/μL), RNase H (NEB, 2.5 U/μL), *E. coli* DNA polymerase (NEB, 10 U/μL).

2.4. cDNA RDA

1. cDNA RDA Buffers and chemicals: 10× *Dpn*II buffer (NEB); 10× ligase buffer (500 m*M* Tris-HCl, pH 7.8; 100 m*M* MgCl$_2$; 100 m*M* DTT; 10 m*M* ATP); 10 m*M* ATP; 5× PCR buffer (335 m*M* Tris-HCl, pH 8.9 at 25°C; 20 m*M* MgCl$_2$; 80 m*M* (NH$_4$)$_2$SO$_4$; 166 μg/mL bovine serum albumin); 4 m*M* dNTPs (4 m*M* each dATP, dCTP, dGTP, and TTP); 10 *M* ammonium acetate (NH$_4$OAc); EE 3× buffer (30 m*M* EPPS [Sigma], pH 8.0 at 20°C; 3 m*M* EDTA); 5 *M* NaCl; tRNA carrier (10 μg/μL); 10× Mung Bean Nuclease buffer (NEB); and 50 m*M* Tris-HCl (pH 8.9).
2. cDNA RDA enzymes: *Dpn*II (NEB, 10 U/μL); T4 DNA ligase (NEB, 400 U/μL); *Taq* DNA polymerase (e.g., GibcoBRL, 5 U/μL); Mung Bean Nuclease (NEB; 10 U/μL).
3. Kits: QIAquick PCR purification kit, including spin columns; binding buffer PB, and wash buffer PE, (Qiagen Ltd).
4. Oligonucleotides R-12 (5'-GATCTGCGGTGA-3'; 1 mg/mL) and R-24 (5'-AGCACTCTCCAGCCTCTCACCGCA-3'; 2 mg/mL); J-12 oligonucleotide (5'-GATCTGTTCATG-3'; 1 mg/mL), J-24 oligonucleotide (5'-ACCGACGTCGACTATCCATGAACA-3'; 2 mg/mL); N-12 oligonucleotide (5'-GATCTTCCCTCG-3'; 1 mg/mL), N-24 oligonucleotide (5'-AGGCAACTGTGCTATCCGAGGGAA-3'; 2 mg/mL). The oligonucleotides should be prepared on a large scale

(1 µM), desalted, purified by high-performance liquid chromatography, and resuspended in water.

2.5. Cloning and Screening

1. 10× *Dpn*II buffer (NEB).
2. 5 mM dNTP mixture (5 mM each of dATP, dCTP, dGTP, and TTP).
3. 10× PCR buffer (GibcoBRL).
4. 50 mM MgCl$_2$.
5. Enzymes: *Dpn*II (NEB, 10 U/µL); *Bam*H1 (NEB 10 U/µL); calf intestinal alkaline phosphatase (1 U/µL); T4 DNA ligase (NEB, 400 U/µL); RNase inhibitor (Promega, 20–40 U/µL); RT (e.g., Superscript II, Life Technologies, 200 U/µL).
6. Kits : QIAquick DNA purification kit (Qiagen Ltd).
7. Plasmids and oligonucleotides: pBluescriptKS+ vector (Stratagene); M13 forward and reverse primers; random hexamer primer (100 pg/µL); oligonucleotide primers tubU, tubL, difference product-specific primers.

3. Methods

3.1. Preparation of Total RNA From <10^7 Cells

1. For suspension cells: Pellet up to 10^7 cells in a Falcon tube by centrifugation at 400g for 5 min at 4°C.
2a. Resuspend the cells (*see* **Note 1**) in Trizol reagent (Life Technologies). Use 0.5 mL of Trizol reagent for up to 10^7 small cells (e.g., lymphocytes). For larger cells (e.g., fibroblasts) use 1 mL for between 5 × 10^6 and 10^7 cells and double the following volumes accordingly.
2b. For adherent cells: Aspirate the medium and, without washing, add 1–2 mL of Trizol reagent per 25-cm^2 flask. Swirl the flask for 5 min or until lysis is complete.
2c. For tissue samples: Rapidly and accurately dissect the required tissues (>100 µg). Homogenize the tissue immediately on ice in 1 mL of Trizol in a 15-mL Falcon tube using a Polytron homogenizer. Increase the volume of Trizol if the tissue sample is greater than 100 µg.
3. Transfer the lysed cells to a 1.5-mL microcentrifuge tube and incubate in Trizol for 5 min at room temperature. Vortex to ensure that the cells are completely disrupted.

4. To each 0.5 mL of cells in Trizol, add 100 μL of chloroform, mix well by shaking for 15 s, and leave at room temperature for 2–3 min before centrifuging at 16,000*g* for 15 min at 4°C.

5. Remove the RNA-containing aqueous upper phase to a new tube, add 400 μL of isopropanol, and incubate at room temperature for 10 min (*see* **Note 2**). Add 1 μL of 10 mg/mL glycogen as a carrier in the precipitation. Centrifuge at 16,000*g* for 20 min at 4°C.

6. Pour off the supernatant, respin for 1 min, and remove the residual supernatant. Wash the pellet with 1 mL of 75% ethanol, vortex, and centrifuge at 12,000*g* for 5 min at 4°C. Remove the ethanol (*see* **Note 3**), and resuspend it in 75 μL of double-deionized, RNase-free water.

3.2. Removal of Residual Genomic DNA

1. Mix in order: 10 μL of 0.1 *M* DTT, 10 μL of 50 m*M* MgCl$_2$, 3.3 μL of 3 *M* sodium acetate, pH 5.3, 0.5 μL of RNase inhibitor, and 1 μL of DNase I (RNase free). Add the mix to 75 μL of total RNA and incubate at 37°C for 15 min.

2. Extract twice with 100 μL of P:C:I. Extract once with C:I and the move aqueous phase to a new tube.

3. Use directly for polyA+ RNA isolation. Alternatively, precipitate by adding 7 μL of 3 *M* sodium acetate and 250 μL of 100% ethanol. Mix and incubate at –20°C for 20 min. Centrifuge at 12,000 rpm for 15 min at 4°C, wash with 70% ethanol, and resuspend the pellet in 20 μL of RNase-free water.

3.3. Preparation of polyA+ RNA

1. Add 100 μL of 2× binding buffer directly to the 100 μL of aqueous phase from the Dnase-treated total RNA.

2. Add 6 μL of oligotex suspension, mix well, and incubate at 65°C for 3 min.

3. Incubate the tube at room temperature for 15 min, mixing gently every 3 min, and prewarm the elution buffer to 75°C.

4. Spin at 16,000*g* for 3 min at room temperature and discard the supernatant (*see* **Note 4**).

5. Wash the oligotex twice in 300 μL of wash buffer, centrifuging for 3 min. Carefully remove the wash buffer.

6. Elute the mRNA from the oligotex with 6 μL of preheated elution buffer by pipetting the Oligotex up and down, vortexing and incubating for 2 min at 75°C.
7. Spin at 16,000g for 3 min at room temperature. Remove the supernatant to a fresh tube and repeat. Combine the supernatants so that the polyA+ RNA is dissolved in 12 μL of elution buffer.

3.4. Preparation of Double-Stranded cDNA

1. Divide the mRNA into two 0.5-mL tubes, one with 10 μL of RNA (+RT), one with 2 μL (–RT; see **Note 5**). Add 10 μL of water to the –RT tube.
2. Incubate the mRNA at 70°C for 5 min, and then place on ice.
3. To each RNA sample add 4 μL of 5× first-strand buffer, 1 μL of 20 mM dNTPs, 2 μL of 0.1 M DTT, 1 μL of 50 ng/μL oligodT primer, 0.5 μL of Rnase inhibitor, and 0.5 μL of Rnase-free water (1.5 μL to the –RT sample). Add 1 μL of RT to the +RT tube and incubate at 42°C for 1 h.
4. Prepare a mixture of 1.8 μL of double-deionized H_2O, 6 μL of 5× RT2 buffer, 0.5 μL of 15 mM βNAD, 0.4 μL of *E. coli* DNA ligase, 0.3 μL of Rnase H, and 1 μL of *E. coli* DNA polymerase.
5. Add 10 μL of the mixture to each tube of first-strand synthesis reaction. Incubate at 15°C for 2 h, and then 22°C for 1 h (see **Notes 6** and **7**).
6. Denature the enzymes used to synthesize cDNA by heating to 70°C for 10 min, and then place the sample on ice.
7. Precipitate the cDNA by adding 1 μL of glycogen carrier (10 mg/mL), 3 μL of 3 M sodium acetate, and 90 μL of 100% ethanol, and then cool to –20°C for 20 min.
8. Centrifuge at 16,000g for 30–60 min at 4°C. Wash the pellet with 70% ethanol and resuspend it in 17 μL of water.

3.5. DpnII Digestion of Double-Stranded cDNA

1. Add 2 μL of 10× *Dpn*II buffer and 1 μL of *Dpn*II to 17 μL of cDNA. Incubate for 1–2 h at 37°C.
2. Denature the *Dpn*II for 20 min at 65°C. Chill on ice and spin briefly to collect the sample at the bottom of the tube.

3.6. Ligation of R-Adaptors

1. Ligate 20 µL of *Dpn*II-digested cDNA to R-adaptors by mixing 31 µL of water, 4 µL of 10× ligase buffer, 2 µL of 10 m*M* ATP, 1 µL of R-24, and 1 µL of R-12 in a 0.5-mL microcentrifuge tube. Add 39 µL of the mixture to 20 µL of digested cDNA.
2. Anneal the oligonucleotides by heating the reaction to 50°C for 1 min and then cooling to 10°C at 1°C/min (*see* **Note 8**).
3. Add 1 µL of T4 DNA ligase, mix, and incubate overnight at 14°C.

3.7. Generation of Representative Amplicons

1. Perform a pilot reaction for each sample. Add 140 µL of water, 40 µL of 5× PCR buffer, 16 µL of 4 m*M* dNTPs, and 1 µL of R-24 primer to a 0.5-mL microfuge tube.
2. Add 2 µL of the ligation reaction to the PCR mix. Place in a PCR machine equipped with a heated lid or overlay with mineral oil.
3. Incubate at 72°C for 3 min (*see* **Note 9**).
4. Pause and add 1 µL of *Taq* DNA polymerase and incubate at 72°C for 5 min.
5. Cycle at 95°C for 1 min, then at 72°C (*see* **Note 10**) for 3 min for various numbers of cycles, finishing with a final extension at 72°C for 10 min.
6. Pause the reaction after 18, 20, 22, and 24 cycles and remove 10-µL aliquots of the product.
7. To each aliquot add 1 µL of 10× agarose gel loading buffer and load onto a 1.5% agarose gel containing ethidium bromide, together with appropriate DNA concentration standards and molecular weight markers. Run the gel.
8. Study the gel and select a cycle number that produces a smear ranging in size from 0.2 to 1.5 kb and which contains approx 0.5 µg of DNA in the 10-µL aliquot.
9. Set up eight reactions for each sample intended for use as driver and two reactions for each sample intended to serve solely as tester.
10. Repeat **steps 2–5**, performing the appropriate number of cycles as determined in the pilot reaction.
11. Combine the eight reactions into two 1.5-mL microcentrifuge tubes (four reactions in each) and extract each with 700 µL of P:C:I, then

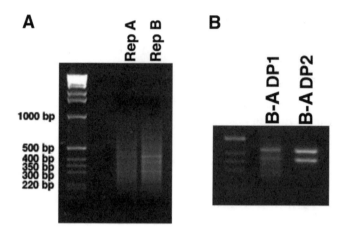

Fig. 3. Agarose gels stained with ethidium bromide showing a typical comparison of two individual representations (**A** and **B**) and a comparison of the difference products, DP1 and DP2.

with 700 μL of C:I. Add 75 μL of 3 *M* sodium acetate (pH 5.3), 750 μL of isopropanol, and precipitate the DNA on ice for 20 min.
12. Centrifuge at 16,000*g* for 15 min at 4°C. Wash the pellet with 0.5 mL of 70% ethanol, dry, and resuspend each of the pellets in 100 μL of TE to give a concentration of 0.5 mg/mL.
13. Combine the products and run 1 μL on a 1.5% agarose gel to check concentration and quality (**Fig. 3**; *see* **Note 11**).

3.8. Generation of the Driver

1. Digest 90 μg (180 μL) of each representation with *Dpn*II. Mix 180 μL of the representation DNA with 340 μL of water, 60 μL of 10× *Dpn*II buffer, and 20 μL of *Dpn*II. Incubate for 2 h at 37°C.
2. Extract with equal volumes of P:C:I, then C:I. Add 60 μL of 3 *M* sodium acetate (pH 5.3) and 600 μL of isopropanol and precipitate on ice for 20 min.
3. Centrifuge the sample at 16,000*g* for 15 min at 4°C. Wash the pellet with 70% ethanol. Dry the pellet and resuspend the digested representation in 180 μL of TE.
4. Assess the concentration by running 1 μL on a 1.5% gel with standards, then dilute it if necessary with TE to 0.5 mg/mL. This is the cut DRIVER.

3.9. Generation of the Tester

1a. In a reciprocal subtraction, a representation will serve as both the tester and driver. In this case, remove 20 μL of the final digested product (see **Subheading 4.8.**) to a separate tube and proceed to **step 4**.

1b. If a representation is to be used only as a tester, digest 10 μg (20 μL) of each representation with *Dpn*II. Mix 20 μL of the representation DNA with 68 μL of water, 10 μL of 10× *Dpn*II buffer, and 2 μL of *Dpn*II. Incubate for 2 h at 37°C.

2. Extract with equal volumes of P:C:I then CLI. Add 10 μL of 3 *M* sodium acetate (pH 5.3), 250 μL of 100% ethanol, and precipitate at –20°C for 20 min.

3. Centrifuge the sample at 16,000*g* for 15 min at 4°C. Wash the pellet with 70% ethanol. Dry the pellet and resuspend the digested representation in 20 μL of TE.

4. Purify the representation away from the digested R-adaptors by using a spin purification column such as QIAquick spin columns.

5. Place a QIAquick spin column in a 2-mL collection tube. Add 100 μL of buffer PB to 20 μLl of digested representation and apply the mixture to the column and allow it to stand for 2 min.

6. Centrifuge the column at 16,000*g* for 30–60 s at room temperature. Discard the flow through.

7. Wash the column with 0.75 mL of buffer PE, centrifuging again for 30–60 s. Discard the flow through and centrifuge for an additional 1 min to remove any residual buffer.

8. Elute the DNA by the addition of 50 μL of TE to the center of the column. Allow the column to stand for 2 min, then centrifuge as before for 1 min.

9. Repeat the elution and combine the eluates to produce 100 μL of cut purified representation in TE.

10. Estimate the concentration of the DNA and the efficiency of adaptor removal by running a 5-μL aliquot next to standards on a 1.5% agarose gel (*see* **Note 12**).

11. Ligate the cut, purified representation to J-adaptors in a 0.5-mL microcentrifuge tube by mixing 1 μg of spin column-purified DNA (usually 10–20 μL) with 3 μL of 10× ligase buffer, 2 μL of 1 mg/mL J-12 oligo, and 2 μL of 2 mg/mL J-24 oligo. Add water to a total volume of 29 μL.

12. Anneal the oligos in a PCR machine by heating to 50°C for 1 min, then cooling to 10°C at 1°C/min.

13. Add 1 µL of T4 DNA ligase and incubate the reaction overnight at 12°C.

14. Dilute the ligation to 10 ng/µL by adding 70 µL of TE. This generates the J-Ligated TESTER.

3.10. Subtractive Hybridization

1. Mix 20 µL (10 µg) of digested DRIVER representation with 10 µL of diluted, J-ligated TESTER representation (0.1 µg) in a 0.5-mL microcentrifuge tube, creating a driver:tester ratio of 100:1.

2. Add 70 µL of water and extract with 100 µL of P:C:I, followed by 100 µL of C:I. Add 25 µL of 10 M NH$_4$OAc, 320 µL of 100% ethanol, and precipitate at –70°C for 10 min.

3. Incubate the tube at 37°C for 1 min, and then centrifuge at 16,000g for 15 min at 4°C. Wash the pellet twice with 70% ethanol, each time spinning at 16,000g for 2 min (*see* **Note 13**).

4. Dry the pellet, and resuspend it very thoroughly in 4 µL of EE 3× buffer by pipetting for at least 2 min, then warming to 37°C for 5 min, vortexing, and spinning to the bottom of the tube.

5. Overlay the solution with 35 µL of mineral oil, place the tube in a PCR machine, and denature the DNA for 5 min at 98°C. Cool the block to 67°C, and immediately add 1 µL of 5 M NaCl directly into the DNA.

6. Incubate the hybridization at 67°C for 20 h.

7. Remove as much oil as possible from the completed hybridization (*see* **Note 14**) and dilute the DNA stepwise in 200 µL of TE. The DNA can be very viscous at this point and should be diluted first by adding 10 µL of TE and repeatedly pipetting, followed by a further 25 µL of TE with more pipetting, and finally made up to 200 µL and vortexed.

8. Store the hybridized DNA at –20°C or use it directly to generate the first difference product.

3.11. Generation of First Difference Product (DP1)

1. For each subtraction, set up two PCRs in 0.5-mL microcentrifuge tubes. For each reaction, mix in order 122 µL of water, 40 µL of 5× PCR buffer, and 16 µL of 4 mM dNTPs. Finally, add 20 µL of the diluted hybridization mix.

2. Place in a PCR machine equipped with a heated lid or overlay with mineral oil.

3. Incubate at 72°C for 3 min.

4. Pause the machine and add 1 µL of *Taq* DNA polymerase to each reaction and incubate at 72°C for 5 min.

5. Just before the 5-min incubation is finished, pause the machine again, and add 1 µL of 2 mg/mL J-24 primer.

6. Cycle at 95°C for 1 min, then at 70°C for 3 min for 11 cycles, finishing with a final extension at 72°C for 10 min (*see* **Note 15**).

7. Combine the two reactions into a single 1.5-mL microcentrifuge tube. Extract with 400 µL of P:C:I and then C:I. Add 10 µL of tRNA carrier, 40 µL of 3 *M* sodium acetate (pH 5.3), 400 µL of isopropanol, and precipitate on ice for 20 min.

8. Centrifuge at 16,000*g* for 15 min at 4°C and wash the pellet with 70% ethanol. Resuspend the pellet in 20 µL of TE.

9. Transfer the 20 µL of DNA to a 0.5-mL microcentrifuge tube and add 4 µL of 10× Mung Bean Nuclease buffer, 14 µL of water, and 2 µL of Mung Bean Nuclease.

10. Incubate the tube in a PCR machine at 30°C for 35 min.

11. Stop the reaction by adding 160 µL of 50 m*M* Tris-HCl (pH 8.9), and incubating at 98°C for 5 min. Place the tube containing the denatured, Mung Bean Nuclease-treated DNA on ice.

12. For each subtraction, set up one PCR reaction in a 0.5-mL microcentrifuge tube by combining 122 µL of water, 40 µL of 5× PCR buffer, 16 µL of 4 m*M* dNTPs, and 1 µL of 2 mg/mL J-24 oligo. Place the tubes on ice.

13. On ice, add 20 µL of the MBN-treated DNA.

14. Place the tubes in a PCR machine and incubate at 95°C for 1 min. Cool the block to 80°C and add 1 µL of *Taq* DNA polymerase.

15. Perform 18 cycles each of 95°C for 1 min then 70°C for 3 min, with a final extension of 72°C for 10 min. Cool the reactions to room temperature.

16. Check a 10-µL sample on a 1.5% agarose gel next to standards to confirm the presence of products and to estimate the concentration.

17. Extract with 200 µL of P:C:I, and then C:I. Add 20 µL of 3 *M* sodium acetate (pH 5.3), 200 µL of isopropanol, and precipitate the products on ice for 20 min.

18. Centrifuge at 16,000*g* for 15 min at 4°C and wash the pellet with 70% ethanol. Resuspend the pellet at 0.5 µg/µL in TE (approx 25 µL, depending on the estimated yield determined in step 5). This is the DP1.

3.12. Change of Adaptors

1. Digest 2 µg (4 µL) of the difference product (from Protocol 13) with *Dpn*II. In a 1.5-mL microcentrifuge tube, mix 4 µL of the DP1 DNA with 84 µL of water, 10 µL of 10× *Dpn*II buffer, and 2 µL of *Dpn*II. Incubate for 2 h at 37°C.
2. Extract with equal volumes of P:C:I then C:I. Add 1 µL of glycogen carrier, 10 µL of 3 *M* sodium acetate (pH 5.3), 250 µL of 100% ethanol, and precipitate at –20°C for 20 min.
3. Centrifuge the sample at 16,000*g* for 15 min at 4°C. Wash the pellet with 70% ethanol. Air-dry the pellet and resuspend the digested representation in 20 µL of TE (100 ng/µL).
4. Ligate 2 µL (200 ng) of the cut difference product (Protocol 14A, step 3) to N-adaptors in a 0.5-mL microcentrifuge tube by adding 3 µL of 10× ligase buffer, 2 µL of 1 mg/mL N-12 oligo, 2 µL of 2 mg/mL N-24 oligo, and 20 µL of water.
5. Anneal the oligos in a PCR machine by heating to 50°C for 1 min, then cooling to 10°C at 1 °C/min. Add 1 µL of T4 DNA ligase and incubate the reaction overnight at 14°C.
6. Add 130 µL of TE to dilute the ligation to 1.25 ng/µL.

3.13. Generation of Second Difference Product (DP2)

1. Mix 10 µL (12.5 ng) of N-ligated DP1 with 20 µL (10 µg) of driver (a driver:tester ratio of 800:1), and proceed through the subtraction hybridization procedure in **Subheading 3.0.**
2. Perform the initial amplification and Mung Bean Nuclease digestion by following the procedure in **Subheading 3.11.**, using N-24 primers in place of J-24, and performing all annealing and extensions during the PCR at 72°C.
3. Generate the final DP2 as in **Subheading 3.11.** Perform two reactions for each sample. Combine the products of the two reactions at **step 17**, in **Subheading 3.11.** and double the volumes of reagents used for extraction and precipitation.
4. Resuspend the pellet from the combined final PCRs at 0.5 µg/µL (usually 50 µL) in TE. This is the DP2 (*see* **Note 16**).

3.14. Primary Screening of Difference Products

1. Digest 12.5 µg of DP2 with *Dpn*II in a total volume of 300 µL. Extract the digest with P:C:I, then C:I, precipitate with ethanol, and resuspend the pellet in 100 µL of TE.

2. Prepare a 1.5% gel with wide lanes and load all 100 µL of cut DP2. Run the gel at a low voltage for several hours to separate the difference products.
3. Cut out each band with a clean scalpel blade and extract the DNA using a QIAquick DNA purification kit (Protocol 11A) or other suitable method.
4. Digest a suitable vector (e.g., pBluescriptKS+) with *Bam*HI. Phosphatase the ends of the vector with calf intestinal alkaline phosphatase and gel purify it (*see* **step 3**).
5. Ligate each difference product into the *Bam*H1 cut vector in a total volume of 5 µL using 50 ng of vector, 10–50 ng of insert, 0.5 µL of 10× ligation buffer, and 0.5 µL of T4 DNA ligase.
6. Transform 3 µL of the ligation into DH5α and grow the bacteria on agar plates containing ampicillin.
7. Prepare a PCR tube for 10 colonies per transformation.
8. Pick the colonies, touch them onto a gridded amp plate to regrow, and place the remainder of the colony into the PCR reactions.
9. Amplify the inserts using the M13 forward and reverse primers.
10. Gel purify products that contain inserts of the predicted size and sequence them directly.
11. Miniprep any interesting products.
12. Generate several representation blots by running 0.75 µg of all the required representations on a 1.5% agarose gel and Southern blotting them onto a nylon membrane.
13. Prepare ^{32}P-labeled probes from the inserts of the difference product plasmids. Hybridize the products to the representation blots to verify true difference products.
14. Screen all true difference products against fresh isolates of RNA.

4. Notes

1. Cell pellets can be effectively disrupted before the addition of medium or Trizol by scraping the base of the tube along an empty Eppendorf rack.
2. Do not leave isopropanol precipitations on ice for more than 1 h at any point in the protocol and never place isopropanol precipitations in a freezer because this produces a heavy salt precipitate that will inhibit later reactions.
3. RNA pellets should be completely free from ethanol but not vacuum dried. This can be achieved with careful use of a drawn out Pasteur pipet attached to an aspirator.

4. The Oligotex pellet is fragile and it is safer not to attempt to take too much of the supernatant. We also suggest that supernatants and washes are respun to ensure material is not lost.

5. Only 20% of the material is used as the negative control to conserve RNA. It is important not to omit this control because it is the only way to assess potential contamination by genomic DNA.

6. OligodT binds to the 3' polyA tail of the mRNA and primes first-strand synthesis. Conditions need to be optimal to obtain the 5' ends of long messages. This can be monitored by including ^{32}P-dCTP in a pilot reaction (or an aliquot), running it on an alkaline agarose gel, and exposing it to film.

7. A negative control (without reverse transcriptase) can be performed to assess potential contamination by genomic DNA.

8. It is convenient to program a PCR machine to perform the gradual cooling.

9. During this period the 12-mer (R-12) dissociates, freeing the 3' ends for fill-in of the ends complementary to the 24-mer adaptors

10. Both primer annealing and extension occur at 72°C.

11. A small quantity of R-adaptors may still be visible, but these should not significantly affect the following ligation because J-adaptors are added in a large excess.

12. It is normal for 5–10 µL to correspond to approx 0.5 µg of cut, purified representation.

13. If the pellet is very difficult to see after the first wash, omit the second wash.

14. Residual oil in the dilution does not impair the subsequent amplification.

15. J-primers have a preferred annealing temperature of 70°C.

16. It is unusual to have to proceed to a DP3. The ratio of driver:tester in DP2 can be varied from 400:1 to 2400:1. This may be preferable to generating a third difference product, where products may be lost.

References

1. Chao, D. T. and Korsmeyer, S. J. (1998) BCL-2 family: regulators of cell death. *Annu. Rev. Immunol.* **16,** 395–419.

2. Vousden, K. H. (2000) p53: death star. *Cell* **103,** 691–694.

3. Rathmell, J. C. and Thompson, C. B. (1999) The central effectors of cell death in the immune system. *Annu. Rev. Immunol.* **17,** 781–828.

4. Hubank, M. and Schatz, D. G. (1994) Identifying differences in mRNA expression by representational difference analysis of cDNA. *Nucleic Acids Res.* **22,** 5640–5648.

5. Hubank, M. and Schatz, D. G. (1999) cDNA representational difference analysis: a sensitive and flexible method for identification of differentially expressed genes, in *Methods in Eneymology cDNA Preparation and Characterization,* Vol. 303 (Weissman, S. M., ed.), Academic Press, San Diego, CA, pp. 325–349.

6. Inukai, T., Inoue, A., Kurosawa, H., Goi, K., Shinjyo, T., Ozawa, K., et al. (1999) SLUG, a ces-1-related zinc finger transcription factor gene with antiapoptotic activity, is a downstream target of the E2A-HLF oncoprotein. *Mol. Cell* **4,** 343–352.

7. Zheng, W. P. and Flavell, R. A. (1997) The transcription factor, G. A.TA-3 is necessary and sufficient for Th2 cytokine gene expression in, C. D.4 T cells. *Cell* **89,** 587–596.

8. Seale, P., Sabourin, L. A., Girgis-Gabardo, A., Mansouri, A., Gruss, P., and Rudnicki, M. A. (2000) Pax7 is required for the specification of myogenic satellite cells. *Cell* **102,** 777–786.

9. Verpy, E., Leibovici, M., Zwaenepoel, I., Liu, X. Z., Gal, A., Salem, N., et al. (2000) A defect in harmonin, a, P. D.Z domain-containing protein expressed in the inner ear sensory hair cells, underlies Usher syndrome type 1C. *Nat. Genet.* **26,** 51–55.

10. Lerner, A., Clayton, L. K., Mizoguchi, E., Ghendler, Y., van Ewijk, W., Koyasu, S., et al. (1996) Cross-linking of T-cell receptors on double-positive thymocytes induces a cytokine-mediated stromal activation process linked to cell death. *EMBO J.* **15,** 5876–5887.

11. Rutherford, M. N., Bayly, G. R., Matthews, B. P., Okuda, T., Dinjens, W. M., Kondoh, H., et al. (2001) The leukemogenic transcription factor E2a-Pbx1 induces expression of the putative N-myc and p53 target gene, N. D.RG1 in Ba/F3 cells. *Leukemia* **15,** 362–370.

12. Liu, H. T., Wang, Y. G., Zhang, Y. M., Song, Q. S., Di, C. H., Chen, G. H., et al. (1999) TFAR19, a novel apoptosis-related gene cloned from human leukemia cell line, T. F.-1, could enhance apoptosis of some tumor cells induced by growth factor withdrawal. *Biochem. Biophys. Res. Commun.* **254,** 203–210.

19

Identification of Apoptosis Regulatory Genes Using Insertional Mutagenesis

Joëlle Thomas, Yann Leverrier, Anne-Laure Mathieu, and Jacqueline Marvel

Summary

This chapter describes a retroviral insertion mutagenesis approach using replication-deficient myeloproliferative sarcoma virus retroviral vectors to identify apoptosis regulatory genes in the interleukin-3-dependent Baf-3 cell line. We describe the retroviral insertion mutagenesis protocol and the selection steps to obtain apoptosis resistant mutants. We also present several methods to isolate the cellular DNA sequences flanking the provirus to identify the gene responsible for the apoptosis-resistant phenotype.

Key Words: Retroviral insertion mutagenesis; retrovirus; apoptosis.

1. Introduction

This chapter describes a retroviral insertion mutagenesis approach to identify apoptosis regulatory genes in the interleukin (IL)-3-dependent Baf-3 cell line. This mutagenesis approach has one major advantage over random mutations induced by drugs or irradiation in that the modified gene can be traced using the provirus (i.e., the integrated form of the retrovirus) as a tag. This was first described by Hayward et al. *(1)*, who showed that slow-transforming retroviruses can both be used as insertional mutagens and as molecular tags allowing the identification of a large number of genes involved

From: *Methods in Molecular Biology, vol. 282: Apoptosis Methods and Protocols*
Edited by: H. J. M. Brady © Humana Press Inc., Totowa, NJ

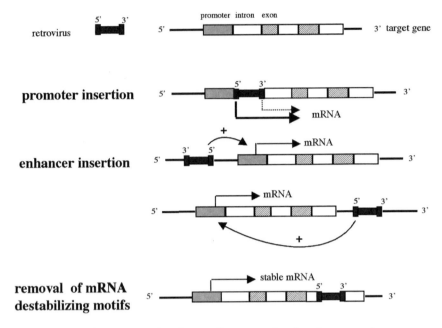

Fig. 1. Different mechanisms of gene activation by retroviruses.

in oncogenesis *(2)*. Random integration of the provirus in the host genome sometimes occurs within or close to cellular genes and potentially deregulates their expression. Different mechanisms of gene activation or inactivation can occur depending on the site of integration and the orientation of the provirus *(2)*. At least three types of mechanisms can be responsible for gene activation (**Fig. 1**): 1) promoter insertion where the viral long terminal repeat (LTR) is used as a promoter for gene expression; 2) enhancer insertion where the LTR acts as an enhancer on the gene promoter; and 3) removal of mRNA-destabilizing motifs when the proviral insertion site is located within the 3' untranslated region of the cellular gene in the same transcriptional orientation. The transcripts are then cleaved at the polyadenylation site of the 5' LTR. This may result in the production of transcripts with increased stability.

Integration of a provirus can also have effects on the protein when inserted into the transcription unit of a gene. This can result in no protein synthesis or in the production of an aberrant protein with no or abnormal biological activity.

Here, we present the retroviral mutagenesis approach using replication-deficient myeloproliferative sarcoma virus (MPSV) retroviral vectors to identify apoptosis regulatory genes in the IL-3-dependent Baf-3 cell line. These cells are grown in culture in the presence of IL-3 and die by apoptosis upon IL-3 removal. The first step after infection is to select cells that have potentially activated an apoptosis inhibitory gene based on their capacity to survive in the absence of IL-3. The expression of several types of genes can allow Baf-3 cells to survive in the absence of IL-3: 1) activation of genes coding for growth factors; 2) activation of genes implicated in the signal transduction mediated by growth factors; 3) activation of apoptosis inhibitory genes; and 4) activation of genes implicated in the regulation of antiapoptotic genes

The second step is to identify the gene responsible for the apoptosis-resistant phenotype. This is based on the identification and isolation of coding sequences located close to the retroviral insertion point. Several strategies will be described in the methods.

2. Materials

2.1. Cell Lines

1. The Baf-3 cell line is derived from murine bone marrow and is dependent on IL-3 for its survival in culture (*3,4*). These cells are grown in suspension and are maintained in DMEM (Life Technologies) containing 6% fetal calf serum (FCS, Roche Molecular Biochemicals), 2 mM L-glutamine (Life Technologies), and 5% WEHI 3B (ATCC TIB 68) cell-conditioned medium as a source of IL-3 (Baf-3 medium; *see* **Note 1**). Cells are grown at a density of 5×10^4 to 10^6 per mL.
2. The PAPM3-adherent cell line producing the M3Pneo-sup retrovirus (*5*) is maintained in DMEM containing 5% newborn calf serum (Life Technologies) and 2 mM L-glutamine.
3. The TEL671 cell line that produces ecotropic pseudotypes of the pMFG nls-LacZ retroviral vector derived from murine leukemia virus (MLV) (*6,7*) is cultivated in Dulbecco's modified eagles medium (DMEM) containing 6% FCS and 2 mM L-glutamine.

A

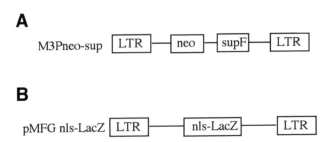

B

Fig. 2. Maps of the integrated provirus. (**A**) M3Pneo sup. (**B**) pMFG nls-LacZ.

All cell lines are grown at 37°C under 7% CO_2 in the presence of 50 µg/mL of gentamycin (Life Technologies).

2.2. Retroviruses

1. M3Pneo-sup containing the neomycin resistance gene and the bacterial tyrosine tRNA suppressor gene (supF; **ref. 5**; **Fig. 2A**).
2. pMFG nls-LacZ expressing the β-galactosidase gene (**ref. 6**; **Fig. 2B**).

3. Methods

3.1. Cells Cloning

Cells are cloned in 1.5% methocel MC (methyl cellulose). 4% methocel is prepared as follows:

1. In a sterile 2-L flask containing a stirrer, boil 150 mL of mineral water (Volvic, France) and add 12 g of methocel MC powder. After the mixture has cooled down, add 150 mL of 2× DMEM (Life Technologies) and mix for 4 d at 4°C.
2. Solubilized nethocel MC is stored at –20°C and supplemented with 6% FCS, 2 m*M* L-glutamine, and 50 µg/mL gentamycin just before use.

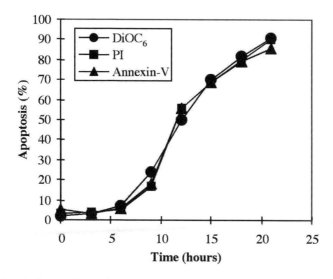

Fig. 3. Detection of apoptotic Baf-3 cells during IL-3 starvation using PI, annexin-V, or $DiOC_6$. Baf-3 cells were starved of IL-3 as described previously *(14)*. The percentage of viable cells was determined at the indicated times by staining with PI, $DiOC_6$, and annexin-V and analysis on a FACscan.

3.2. Apoptosis Measurement

The percentage of apoptic cells is measured by propidium iodide (PI) exclusion and analyzed on a FACscan (Becton-Dickinson). To measure the percentage of apoptotic cells, PI (Sigma) is added to an aliquot of Baf-3 cells at a final concentration of 5 μg/mL. After a 5-min incubation cells are analyzed on a FACscan using a FSC/SSC gate to exclude debris. For each time point at least 3000 cells are analyzed on FSC/FL2 dot plot. This staining protocol allows the distinction between late apoptic (FL-2 bright, FSC low), apoptotic (FL-2 dull, FSC intermediate), and live cells (FL-2 negative, FSC high; **ref. 8**).

Apoptosis can also be measured using annexin-V labeling (Roche Molecular Biochemicals) or $DiOC_6$ (Molecular Probes), which are both markers of early stages of apoptosis. In our hands, all three methods give similar results using Baf-3 cells (**Fig. 3**).

Hence we used PI exclusion because this approach allows measurement of the percentage of apoptotic Baf-3 cells in the culture almost instantaneously.

3.3. Retroviral Infection

3.3.1. Infection With M3Pneo-sup by Coculture

5×10^5 to 2×10^6 Baf-3 cells in 10 mL of Baf-3 medium supplemented with 8 µg/mL of Polybrene (Sigma) and 5% newborn calf serum are added to a 10-mm culture dish seeded with semiconfluent M3Pneo-sup producing cells (PAPM3 cells) for 48 h. Usually, 6 to 7×10^7 Baf-3 cells are incubated with the retrovirus producing cells per mutagenesis experiment (*see* **Note 2**).

3.3.2. Infection With pMFG nls-LacZ Using Viral Supernatant

1. Add one volume of viral supernatant of pMFG nls-LacZ-producing cells containing 10^7 infectious particles per milliliter to three volumes of Baf-3 cells at a density of 2×10^5 cells/mL (use a total of 6 to 7×10^7 Baf-3 cells) and culture for 4 h in Baf-3 medium supplemented with 8 µg/mL of Polybrene.
2. After 4 h, remove medium and wash cells once with Baf-3 medium.
3. Add fresh Baf-3 medium and culture cells for another 2 d to allow provirus integration and the resulting expression or modification of neighboring cellular genes to occur (*see* **Note 2**).

3.4. Selection of Apoptosis-Resistant Baf-3 Clones

1. Two days after infection, harvest Baf-3 cells by centrifugation and wash cells twice with Baf-3 medium without IL-3.
2. Grow cells in Baf-3 medium without IL-3 until more than 95% of the cells are dead (approx 30 h; *see* **Note 3**).
3. Eliminate dead cells by Ficoll density gradient according to the manufacturer protocol (Lympholyte-M, Cedarlane laboratories, Hornby, Canada).
4. The surviving cells are cloned in the absence of IL-3 in methocel MC 1.5% prepared as described in materials (*see* **Notes 4** and **5**).
5. After 30 h of culture in methocel, add 5% IL-3 back to the dishes to allow the surviving cells to recover (*see* **Note 3** for the determination of the starvation duration time).

6. After 10 d of culture, select individual clones and expand them in liquid Baf-3 medium.
7. Before molecular analysis, each clone is individually tested for its resistance to apoptosis induced by IL-3 starvation. Cell viability is generally measured 20 h after IL-3 removal by PI exclusion.
8. Clones that are strongly resistant to apoptosis are recloned by limiting dilution (**ref. 9**; *see* **Note 6**).

3.5. Molecular Analysis of Apoptosis-Resistant Baf-3 Clones

1. Analyze for expression of known antiapoptotic genes using classical Northern and Western blots techniques.
2. Analysis of Southern blots is realized in order to verify that only one retroviral integration has occurred in the selected mutants *see* **Note 7**).

3.6. Isolation of Cellular DNA in the Vicinity of the Retroviral Insertion Point

We will describe several methods that we have used successfully to identify cellular DNA sequences flanking the provirus. Recently other methods have been developed such as linear amplification-mediated polymerase chain reaction (LAM-PCR; **ref. 10**).

3.6.1. Cloning

This method is based on the selection of clones containing proviral sequences from a subgenomic DNA library. These libraries are constructed with DNA fragments of a size corresponding to the one hybridizing with the provirus. Hence they are enriched in DNA fragments containing the integrated provirus allowing the screen of a smaller number of clones.

1. Digest 60 μg of genomic DNA with restriction enzymes that do not cut in the provirus, that is, *Eco*RI for mutants that have been obtained with the M3Pneo-sup vector.
2. Load a 0.8% agarose gel with 10 and 50 μg of digested DNA.
3. After migration cut the gel in half, store the part of the gel containing the 50 μg of DNA (preparative gel) at 4°C, and incubate the second part of the gel containing the 10 μg of DNA in a 1 μg/mL EtBr solution for 30 min.

4. Take a picture of the gel in the presence of a ruler.

5. Transfer DNA on a N$^+$ nylon membrane (Amersham) and hybridize with a probe specific for the vector (i.e., neomycin for the M3Pneo-sup vector)

6. Localize precisely the DNA fragment hybridized to the probe and cut the corresponding band from the preparative gel stored at 4°C.

7. Purify DNA from the gel using GlassMax (Life Technologies) or equivalent.

8. Choose an appropriate cloning vector depending on the size of the purified DNA fragment to generate a subgenomic library enriched in DNA fragments containing the provirus.

9. Screen the subgenomic library with a probe specific for the retroviral vector (i.e., neomycin for the M3Pneo-sup vector).

10. Purify and sequence the positive clone to characterize the cellular sequences flanking the provirus.

11. This method can be combined with the GenomeWalker (**Subheading 3.6.3.**) to isolate more cellular DNA.

3.6.2. Inverse PCR

Inverse PCR can be used to recover DNA fragments flanking a known DNA sequence (i.e., in our case the sequence of the integrated provirus; **Fig. 4**; **ref. *11***). The success of this approach depends a lot on the choice of the primers. It is best to design primers from regions specific to the retrovirus, for example, from the neomycin gene for M3Pneosup or the *LacZ* gene in the case of pMFG nls-LacZ. This also allows amplification of some of the provirus genome that can easily be identified in the sequenced PCR product.

1. Digest 10 µg of genomic DNA with different restriction enzymes.

2. Purify DNA by phenol/chloroform extraction and ethanol precipitation.

3. Resuspend the DNA pellet in 10 µL of H$_2$O.

4. Set up a ligation reaction with 10 ng of DNA. The low DNA concentration in the ligation reaction will favor circularization of the DNA molecules.

5. Ligation is realized in a 50 µL final volume with 2 units of ligase (Life Technologies) for 16 h at 16°C.

6. Inactivate ligase 15 min at 65°C.

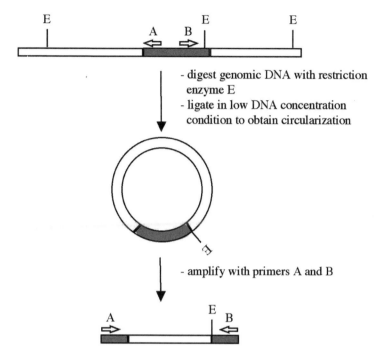

Fig. 4. Principle of inverse PCR.

7. Precipitate DNA with ethanol.
8. Resuspend the DNA pellet in 10 µL of H$_2$O.
9. Using 3 µL of ligated DNA set up a long template PCR following the supplier's instructions (Roche Molecular Biochemicals).
10. Set up a second long template PCR on 1 µL of PCR product using internal primers (nested PCR).
11. Analyze 10 µL of PCR product on a 0.7% agarose gel.
12. Purify and sequence the PCR product and check for the presence of provirus sequences to make sure the right PCR product has been isolated.

3.6.3. GenomeWalker

The method mentioned here is an alternative to the inverse PCR. We have used the GenomeWalker kit from Clontech (*see* **Note 8**). The principle of this method is described in **Fig. 5**.

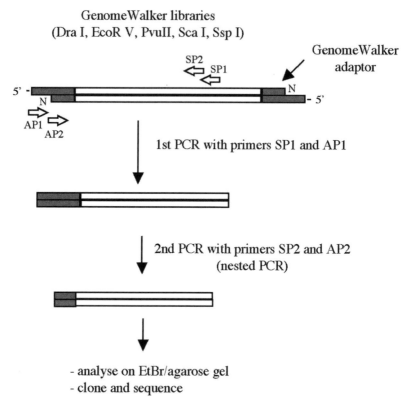

Fig. 5. Principle of the GenomeWalker kit from clontech. SP1 and SP2 are primers specific for the retroviral vector. AP1 and AP2 are adaptor primers provided in the kit. N is an amine group that blocks extension of the 3' end of the adaptor-ligated genomic fragments.

3.7. Analysis of the Cellular Sequences Isolated in the Vicinity of the Provirus

Once the cellular DNA flanking the provirus has been isolated, different analyses can be performed.

1. The identity of the cellular DNA isolated by these techniques can be confirmed by Southern analysis. Indeed a retrovirus-induced rearrangement should be found in the mutant but not in the parental cells using the cellular DNA fragments as a probe *(7)*.

2. After the DNA has been sequenced, homologous sequences are searched for in nucleotide or protein databases. If no homologous sequence is found, the sequence can be further analyzed for the presence of potential open reading frames or noncoding exons. If potential exons are identified probes can be derived to check by Northern blot analysis the overexpression of the corresponding gene in the mutant (*see* **Note 9**).

4. Notes

1. Expand WEHI 3B cells in DMEM medium containing 6% FCS and 2 m*M* L-glutamine, maintaining the cell concentration between 2×10^4 and 2×10^5 cells/mL until desired volume is obtained.
   - Let the cells reach confluence (1×10^6 cells/mL) and before too many (1 to 5%) dead cells appear (after 24 to 48 h), harvest the supernatant.
   - Remove cells by centrifugation and filter supernatant on a 0.22-μm filter.
   - Freeze aliquots of the IL-3 supernatant at –20°C.
   - For each batch of supernatant produced, the optimal concentration to be used for Baf-3 culture is determined by proliferation and survival assays.

2. The choice between different infection modes will vary according to the target cell type and the retroviral packaging cell line. Coculture is in general favored when viral titers are low in the supernatant. However, it has been shown recently that viral infection efficiency can be enhanced by spinoculation, that is, cells are centrifuged with retroviral supernatant at 1200*g* for 2 h at 25°C *(12)*. Moreover, viral supernatants can also be concentrated by centrifugation at 10,000*g* for a few hours at 4°C. The top fraction (corresponding to approximately two thirds of the total volume) is removed and discarded and the bottom fraction is used for infection *(13)*. According to the retroviral construction used for the mutagenesis, different assays can be used to determine the viral infection efficiency. With the M3Pneosup vector, the resistance to the G418 is used to measure the transfection efficiency. This is performed by determining the frequency of infected cells growing under limiting dilution *(9)* in the presence or absence of G418. In the case of cells infected with the pMFG nls-LacZ, an aliquot of infected cells are stained using the β-galactosidase

substrate CMFDG according to the manufacturer's instructions (DectaGene Green Expression Kit, Molecular Probes). The percentage of β-galactosidase-positive cells is determined by analysis on the FACscan (Becton-Dickinson).

3. Selection of mutants by growth factor starvation is a powerful approach. However, overexpression of different apoptosis inhibitory genes will delay the onset of death for variable length of time. Hence, according to the duration of growth factor starvation used different types of genes could be identified. Therefore, it is important to use a growth factor starvation time which will result in the elimination of almost all parental cells while allowing the survival of cells expressing a given apoptosis inhibitory gene. In our case, we have used Baf-3 cells overexpressing the anti-apoptotic gene Bcl-2 to optimize the selection protocol for the identification of Bcl-2-like genes *(14,15)*.

4. As an alternative to methocel cloning, one can also select growth factor starvation-resistant cells by culture under limiting dilution in the absence of IL-3 *(9)*. The growth factor is added to the plates after a given starvation time (*see* **Note 3**).

5. In the case of cells infected with the M3Pneo-sup retrovirus, if the growth factor selection is not stringent enough, G418 (Life Technologies) at 2 mg/mL can be added during the whole methocel cloning step. However, one should be cautious because this could result in the elimination of clones with solo LTR insertion (*see* **Note 7**).

6. Clones that have been isolated by growth factor starvation can be further characterized by testing their resistance to different apoptosis-inducing pathways. For example the use of staurosporine at 0.2 µg/mL allows the distinction between cells overexpressing genes from the Bcl-2 family and cells producing their own growth factors *(7)*. Indeed, cells overexpressing Bcl-2 family antiapoptotic genes are resistant to staurosporine, whereas IL-3 is not efficient enough to protect Baf-3 cells against apoptosis induced by staurosporine.

7. Before starting to identify the activated gene using the provirus as a tag, one has to make sure that there is only one integration in the cellular genome. This is easily performed by Southern blot analysis of genomic DNA extracted from the mutants and by using a probe specific for the retroviral vector used for infection. We have also noticed that in some cases integration of solo LTR can occur. Some of these solo LTRs can lead to strong gene activation *(7)*. To detect solo LTRs, hybridize the Southern blot with a probe specific of LTR derived from MPSV retrovirus. We have used a 220-bp PCR frag-

ment obtained with the 5' primer 5'-GTAGGTTTGGCAAGC-3' and the 3' primer 5'-CAGGAACTGCTTACC-3', both of which hybridize in the U3 region of the MPSV derived LTR. This stretch of MPSV has been chosen because of its low homology to the MuLV U3 region. As a consequence, it allows the specific detection of MPSV LTR because it only marginally hybridizes with endogenous LTR *(7)*.

8. Clontech has now developed a universal GenomeWalker kit, which allows the construction of libraries from the genome of any species.
9. During oncogene identification by insertional mutagenesis, some cases have been reported where no gene close to the integration point could be isolated *(16)*. In such situations, the integrated provirus cannot be used as a molecular tag to identify nearby genes, and other methods such, as differential display PCR, can be used to identify the activated gene.

Acknowledgments

We thank Patrice Dubois and François-Loïc Cosset for their critical reading of the manuscript. Anne-Laure Mathieu is supported by fellowship from the Centre National de la Recherche Scientifique. The work performed in Jacqueline Marvel's laboratory is supported by institutional grants from INSERM and the Ministère de l'Enseignement Supérieur et de la Recherche and by additional support from the Association pour la Recherche sur le Cancer, the Région Rhône-Alpes, and the Comité Départemental du Rhône de la Ligue Nationale Française Contre le Cancer.

References

1. Hayward, W. S., Neel, B. G., and Astrin, S. M. (1981) Activation of a cellular oncogene by promoter insertion in ALV-induced lymphoid leukosis. *Nature* **290,** 475–480.
2. Jonkers, J. and Berns, A. (1996) Retroviral insertional mutagenesis as a strategy to identify cancer genes. *Biochim. Biophys. Acta* **1287,** 29–57.
3. Palacios, R. and Steinmetz, M. (1985) IL3-dependent mouse clones that express B-220 surface antigen, contain Ig genes in germ-line configuration, and generate B lymphocytes in vivo. *Cell* **41,** 727–734.

4. Rodriguez-Tarduchy, G., Collins, M., and Lopez-Rivas, A. (1990) Regulation of apoptosis in interleukin-3-dependent hemopoietic cells by interleukin-3 and calcium ionophores. *EMBO J.* **9,** 2997–3002.

5. Stocking, C., Bergholz, U., Friel, J., Klingler, K., Wagener, T., Starke, C., et al. (1993) Distinct classes of factor-independent mutants can be isolated after retroviral mutagenesis of a human myeloid stem cell line. *Growth Factors* **8,** 197–209.

6. Cosset, F. L., Morling, F. J., Takeuchi, Y., Weiss, R. A., Collins, M. K., and Russell, S. J. (1995) Retroviral retargeting by envelopes expressing an N-terminal binding domain. *J. Virol.* **69,** 6314–6322

7. Thomas, J., Leverrier, Y., and Marvel, J. (1998) *Bcl-X* is the major pleiotropic anti-apoptotic gene activated by retroviral insertion mutagenesis in an IL-3 dependent bone marrow derived cell line. *Oncogene* **16,** 1399–1408

8. Leverrier, Y., Thomas, J., Perkins, G. R., Mangeney, M., Collins, M. K., and Marvel, J. (1997) In bone marrow derived Baf-3 cells, inhibition of apoptosis by IL-3 is mediated by two independent pathways. *Oncogene* **14,** 425–430

9. Henry, C., Marbrook, J., Vann, D. C., Kodlin, D., and Wofsy, C. (1980) Limiting dilution analysis, in *Selected Methods in Cellular Immunology* (Mishell, B. B., and Shiigi, S. M., eds), Freeman and Company, New York, NY, pp. 138–152.

10. Schmidt, M., Zickler, P., Hoffmann, G., Haas, S., Wissler, M., Muessig, A., et al. (2002) Polyclonal long-term repopulating stem cell clones in a primate model. *Blood* **100,** 2737–2743.

11. Silver, J. and Keerikatte, V. (1989) Novel use of polymerase chain reaction to amplify cellular DNA adjacent to an integrated provirus. *J. Virol.* **63,** 1924–1928.

12. O'Doherty, U., Swiggard, W. J., and Malim, M. H. (2000) Human immunodeficiency virus type 1 spinoculation enhances infection through virus binding. *J. Virol.* **74,** 10074–10080.

13. Yang, J., Friedman, M. S., Bian, H., Crofford, L. J., Roessler, B., and McDonagh, K. T. (2002) Highly efficient genetic transduction of primary human synoviocytes with concentrated retroviral supernatant. *Arthritis Res.* **4,** 215–219

14. Leverrier, Y., Thomas, J., Mangeney, M., and Marvel, J. (1996) Characterisation of a retroviral insertion mutagenesis protocol to obtain mutants with activated apoptosis inhibitory genes, in *Tumor*

Biology (Tsiftsoglou, A. S., Sartorelli, A. C., Housman, D. E., and Dexter, T. M., eds), Springler-Verlag, Berlin, Germany, pp. 176–180.

15. Marvel, J., Perkins, G. R., Lopez Rivas, A., and Collins, M. K. (1994) Growth factor starvation of bcl-2 overexpressing murine bone marrow cells induced refractoriness to IL-3 stimulation of proliferation. *Oncogene* **9,** 1117–1122

16. van Lohuizen, M., and Berns, A. (1990) Tumorigenesis by slow-transforming retroviruses: an update. *Biochim. Biophys. Acta* **1032,** 213–235.

Index